The forces of nature

P. C. W. DAVIES
LECTURER IN APPLIED MATHEMATICS
KING'S COLLEGE LONDON

CAMBRIDGE UNIVERSITY PRESS
CAMBRIDGE
LONDON · NEW YORK · MELBOURNE

Published by the Syndics of the Cambridge University Press
The Pitt Building, Trumpington Street, Cambridge CB2 1RP
Bentley House, 200 Euston Road, London NW1 2DB
32 East 57th Street, New York, NY 10022, USA
296 Beaconsfield Parade, Middle Park, Melbourne 3206, Australia

First published 1979

Text set in 10/12 pt VIP Times, printed and bound
in Great Britain at The Pitman Press, Bath

Library of Congress Cataloguing in Publication Data

Davies, PCW
The forces of nature.

1. Nuclear forces (Physics) 2. Nuclear physics.
I. Title.
QC793.3.B5D38 539.7 78–72084
ISBN 0 521 22523 X
ISBN 0 521 29535 1 pbk

Contents

Preface

The purpose of this book is to explain the important concepts underlying the spectacular advances made recently in our understanding of the microcosmos – the world within the atom. It is a world inhabited by a fascinating collection of particles and forces and ruled by laws, some strange, some familiar. Although in many ways it resembles a scaled-down version of the everyday world, bizarre phenomena can occur that have no counterpart in the experience of our senses. When we probe into the organization of the subatomic domain, we must be prepared for situations and circumstances totally beyond our imagination, though not, one hopes, beyond our comprehension.

Very few books exist which attempt to explain modern subatomic physics at the conceptual level. The subject of quantum mechanics, in particular, is almost exclusively confined to formal textbooks. This book explains the basics of subatomic particles and their interactions without subjecting the reader to the tedium of punishing mathematics, or formal sets of rules. The emphasis is on concepts rather than facts and figures, on understanding rather than knowing.

I have written in a style which attempts to communicate my own feelings for the subject: that it is not just a dry accumulation of data, but a human adventure without precedent. I regard subatomic physics as a journey into a parallel universe – a world which co-exists inside us, and all of matter. A world full of strange inhabitants and chequered relationships, where different species of matter appear, play out their roles, and disappear. The quest to find out what the world is made of, and to uncover the forces that lie at the heart of matter, must rank as one of the most significant enterprises of mankind.

But there is more to subatomic physics than the desire to know how the microcosmos is arranged and managed. It is also the testing ground for the two great revolutionary theories of twentieth-century science: relativity and quantum theory. Both enter into the affairs of elementary matter in a fundamental way, and the success of the revolutionary marriage has been stunning.

I have concentrated on the quantum theory, first because it is conceptually harder than relativity, but also because I have treated relativity in detail in my companion book *Space and time in the modern universe*. Both books are written at the same level and style. The level corresponds roughly to that of *Scientific American* or *New Scientist* and I hope will appeal to a wide range of readers, from the off-duty, non-specialist scientist, through students requiring a foundation in the topics of atomic, nuclear and elementary particle physics, quantum mechanics and field theory, to sixth formers and laymen who wish to extend their knowledge of science beyond the occasional snippet in the daily press.

Although many Greek symbols are used as names for particles, there is very little actual mathematics in this book, and none beyond ordinary secondary school algebra and arithmetic. SI units are used throughout except where subatomic distances or particle masses are concerned. There we use the more familiar fermi (10^{-15} m), abbreviated fm, for distance; masses are based on the unit MeV/c^2, or millions of electron volts ÷ (speed of light)^2, which will be abbreviated MeV.

I should like to thank Professor J. G. Taylor for helpful discussions, and Mrs M. Woodcock for assistance in preparation of the manuscript.

August 1978 P.C.W.D.

1
Forces and fields

Most scientists are about 2 m tall, and their apparatus rarely more than a couple of orders of magnitude larger or smaller. Experimental science began by exploration of the world on this sort of scale, but the imagination of its practitioners knew no such dimensional confinement, and they enthusiastically speculated on the structure of the world from the atom to the universe. Advances in technology (e.g. telescopes, atom smashers) have enabled these speculations to be tested and fashioned into full-blown theories.

It was natural enough that the first attempts to describe both the very large and very small should simply involve *extrapolations* of the principles discovered about the world of our daily experience. The only considered difference in the behaviour of atoms and, say, billiard balls, was one of scale. In this century it has been discovered that atoms possess qualities that are totally absent in the macroscopic domain. Indeed, the microcosmos as a whole seems to be ruled by principles, some of which are quite alien to our intuition.

Fortunately for our comprehension there are features about subatomic science that are recognizable and familiar: electricity, for instance. The forces of nature which operate in the microworld can only be understood by first grasping the laws of the macroscopic world, the principles which govern the motion of bodies and the way the forces of gravity and electromagnetism control the objects that we see about us. In this chapter the essential ideas of forces and motion will be sketched. The treatment is intended only as an introduction, to enable the reader to comprehend fully the later sections about the subatomic world itself.

1.1 *Action and motion*

In nature things move violently to their place and calmly in their place (Francis Bacon, 1561–1626)

Why do things happen? For centuries this most fundamental of all questions about the physical world has lain at the foundation of science. An explanation for events must rest on a correct understanding of the natural condition of objects. The world is full of objects – people, planets, clouds, atoms, flowers – and full of motion. Things happen when moving objects act collectively. How do objects *know* about each other? How do they respond to the presence and activities of other objects?

Before the seventeenth century the natural state of all objects on Earth was thought to be the state of *rest* (see the quotation above). In daily experience moving objects soon return to rest if they cease to be powered by something. Switch off the engine of a motor car and it grinds to a halt in a few hundred metres. People did not ask: 'Why does a body remain at rest?' Instead they asked: 'Why does a body move?' The Greek philosopher Aristotle (384–322 B.C.) taught that all motion requires a *cause*, an agency to produce it. For example, the flight of an arrow is sustained by a vortex of air surrounding it. Without an agency to sustain it, all motion naturally ceases. For Aristotle, any activity implied *action*.

Aristotle's ideas were a failure. A proper understanding of motion, action and activity had to await the epochal work of Isaac Newton (English, 1642–1727). Central to Newton's concept of motion is that the state of rest is only one of many possible natural states for an object. A body can also *move uniformly* without the necessity of any external agency or the action of any influence to sustain it. That is, it can, and indeed will, continue to move in a straight path with undiminished speed unless something operates to change it. For Newton, there can be activity (of a limited sort) without action.

The idea that uniform motion does not require any explanation is intimately bound up with the notion that such motion is merely *relative* to whoever observes it. The relativity of motion will be further discussed in Section 1.5, but for now we simply note that the bowl of soup which is at rest on the table before the airplane passenger is travelling at several hundred kilometres per hour over the head of the city dweller below. The 'natural' state of rest for the passenger is a 'natural' state of steady motion for the city dweller.

If uniform, undiminished motion is the natural order of things, why do we need engines for our motor cars? Newton's ideas say that the motion itself of the freewheeling car does not require explanation. Instead we must explain why it slows down to rest. The explanation is readily available. Friction and air resistance sap the energy from the moving car and slow it down. In outer space, where there is no air, and negligible friction, objects continue to move without propulsion. This is fortunate, otherwise the unpowered Earth would stop in its orbit around the sun.

If uniform, unchanging motion is natural and needs no explanation, physics is charged with the task of explaining how bodies are disturbed from this natural state. *Changes* in motion require the action of some external agency to initiate and sustain them. Our everyday world abounds with agencies that act on stationary or uniformly moving bodies to redirect, accelerate or decelerate their motion.

Because the state of uniform motion is regarded as natural, we say that when a body is disturbed from this state it is being *forced*. The agencies which produce forced motion are called forces. It is the action of forces which enriches the activity of our universe, and which enables different parts of the world to be aware of each other's existence. Without forces, nothing could act on or influence anything else, and all the matter in the universe would disintegrate into its elementary constituents, each subatomic particle moving independently of all the others.

Once we are aware of the importance of forces two tasks present themselves. One is to understand the consequences of the action of a force on the bodies which are disturbed by it. The second is to understand the nature and origin of the force itself, and thereby perhaps to learn to control it and direct it through our technology. Newton first demonstrated the consequences of forces through his celebrated second law of dynamics. It is hard to overemphasize the importance of this law, which lies at the very foundation of science. Upon it rests much of what we regard as commonplace engineering. From the stability of a bridge, through the mechanism of a watch, to the motion of a bullet or spacecraft, whenever one wants to describe how mechanical things happen, Newton's second law is invoked.

As always in physical science it is our ability to be able to use simple *mathematical* relationships between physical quantities that makes the scientific method so successful. We can then examine mathematical models of the real world to find out what will happen to systems which we have never even seen or to which we have no access. It was through the mathematical formulation of this second law that Newton was able to describe correctly the shapes and sizes of the planetary orbits millions of kilometres away. Mathematics is the secret of scientific success.

The familiar statement of the second law is simply that the action of a force F produces on a body an acceleration a in proportion to the magnitude of F and in the same direction as F acts. However, a given force will accelerate a heavy body more slowly than a light one, and this is taken into account by introducing the idea of *inertia*. The inertia of a body is a measure of its resistance to forces, and roughly speaking one can think of more massive bodies (motor cars) as having a greater inertia than less massive ones (bicycles). It is easier to push a bicycle than a car. If we denote the mass by m then the second law states

$$F = ma \tag{1.1}$$

or in words, force = mass × acceleration.

To determine how a body responds to a given force F, which may be varying from time to time and place to place in both magnitude and direction, it is necessary to solve Equation (1.1) for the *position* of the body. This can be done because the acceleration a is the rate at which the velocity of the body changes with time and the velocity is, in turn, the rate of change of distance of the body from some fixed place. Solving Equation (1.1) therefore amounts to determining how the *location* of the body changes from time to time, i.e. it determines the *trajectory* or *orbit* which the body follows. That is what is meant by finding out what happens to the body under the action of the force F.

In this book we shall be more interested in the nature of the force itself rather than its consequences, although the two cannot always be separated. Below are listed 12 examples of familiar forces. At the time of Newton very little was known about the nature and origin of forces. Of course, this did not prevent the solution of Equation (1.1) for a large variety of important and interesting physical systems, for it is not necessary to know what causes a force to enable its consequences to be evaluated. Since then, the remarkable fact has emerged that all the forces listed are really manifestations of just *two* fundamental types of force: gravity and electromagnetism. This century two more forces have been discovered but these operate mainly over subatomic distances, and their discussion will be deferred until later chapters.

Pull of a rope	Lightning strike
Blow of a hammer	Raising of the tides
Gravity	TNT explosion
Friction	Magnetism
Air resistance	Wind power
Muscle power	Hydraulic pressure

Examination of these examples shows that the forces listed divide into two qualitatively dissimilar categories. Some, like the blow of a hammer, act by direct contact with the body concerned while others, like the pull of the Earth's gravity or the repulsion between magnets, seem to reach out across empty space. Paradoxically, it is easier to understand the latter action-at-a-distance than action by contact. The reason for this is the obscurity surrounding exactly what is meant by the contact of material bodies. When a golf club strikes a golf ball and delivers a sudden change to the ball's state of rest, we have an intuitive picture of a brief interval during which the head of the club is actually 'touching' the ball and accelerating it. But to comprehend what touching means requires a knowledge of the detailed structure of the materials from which the club and ball are made. As we shall see, the notion of direct contact fails completely at the atomic level.

On the other hand, the concept of action-at-a-distance is a hard one to grasp. When two magnets try to push each other apart it is clear that one magnet somehow 'knows' of the other's existence even though they are separated by an empty gap. How is this knowledge obtained? What sort of influence or message travels across the space between them to power the repulsion? Gravity operates similarly, across millions of kilometres of space, to bind the planets of the solar system to the sun and to raise the ocean tides by the action of the moon. How can sea-water 'know' the position of the moon?

In modern physical theory we have a very intuitive model in terms of field quanta of the way in which the message gets through, which will be described in Chapter 4. At the time of Newton, however, the mechanism of action-at-a-distance was a complete mystery. Newton himself was the first to develop a viable theory of gravity-at-a-distance, without the necessity of explaining how it worked. He simply provided a highly successful mathematical formula which describes the strength of the gravitational force between two material bodies, and which was sufficient to enable him to calculate correctly the regularities of the planetary orbits. Newton's theory of gravity has only been replaced this century, and even today remains almost universally an adequate approximation.

Before describing various specific forces, such as gravity, we shall first note some very fundamental properties about the motion of material bodies which will prove of vital importance in the coming chapters.

First, it is obvious that all forces have both a magnitude (weak, strong) and a direction (e.g. north, downwards). Quantities that can be directed are called *vectors*; we shall meet several examples of vectors in this book. They can be conveniently denoted by arrows. Vectors can be added, but the rules of addition are more complicated than those for numbers. For example, if we add two vectors of equal magnitude but opposite direction (called antiparallel) they cancel each other completely (see Fig. 1.1). On the other hand, if they are parallel, they reinforce to produce a vector in the same direction but with twice the magnitude. Intermediate cases produce vectors with intermediate magnitudes *and* directions. Forces are obviously like this: just think of a tug of war as an example of an antiparallel vector.

Other examples of vectors are velocity, momentum and acceleration. They are all pointed in certain directions. In contrast, temperature or energy have numerical values at a place, but are not directed anywhere. They are not vectors. Newton's second law relates force to acceleration, so it is a vector equation. It gives information not only about the magnitude of the acceleration that a given force produces, but also about its direction – the body is accelerated along the line of direction of the force.

(*a*)

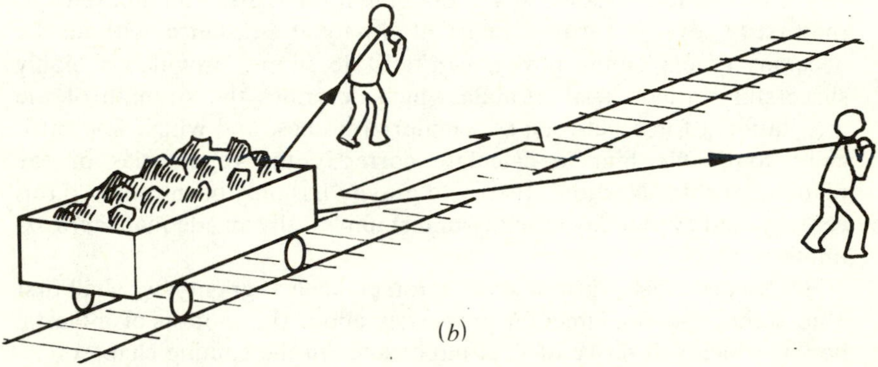

(*b*)

Fig. 1.1. Vectors. Quantities such as force and acceleration have both direction and magnitude: they are called vectors. When vectors are added, account must be taken not only of how large they are, but also which way they point. In the example shown in (*a*) the directions of the forces are opposed and cancel each other, even though their individual magnitudes may be large. In (*b*) obliquely directed forces combine to give a net force in a direction between them with a strength somewhat less than their combined magnitude.

When a force acts on a body and it accelerates, its *momentum* changes. Momentum is a familiar property in daily life. Heavy, fast-moving bodies carry a lot of momentum; light, slow bodies very little. Physicists make this concept precise by defining momentum to be the product of mass (m) and velocity (v): momentum = mv. Now acceleration is simply the *rate* at which velocity changes, so if the mass m remains constant, Equation (1.1) can be written

$$\text{force} = \text{rate of change of momentum.} \qquad (1.2)$$

Even if m varies, this equation remains correct.

Because velocity is a vector, a body can accelerate by changing either the magnitude of its velocity, i.e. its *speed*, or its direction, or both. When a football is kicked from rest it accelerates by changing its speed only. When a bullet ricochets off a target it changes its direction and loses speed. If a spacecraft orbits around the Earth in a circle, it does not alter its speed, but it still accelerates continually by changing the *direction* of its motion. All these types of motion alter the momentum of the body and all require a force to produce it.

If no force acts on a body its momentum will not change. A physicist says that it is *conserved*. The conservation of momentum plays such an important part in the behaviour of subatomic matter that its meaning should be grasped by the reader at this stage. Momentum may be added like any other vector. The total momentum of a system is the sum of the momenta of its constituent parts. If a system is isolated from external forces, its total momentum will not change, but it may still experience internal forces which can rearrange the total momentum among the components. The law of conservation of momentum tells us that the vector sum of all this momentum cannot increase or decrease.

A good example is provided by the gun. At rest, it has zero momentum, so provided it is not struck by something the combined system gun + bullet must have zero momentum also after the bullet has been fired. The force which propels the bullet is purely internal to the system and reacts with an equal force back on the gun. The outcome is a *recoil*. The momentum carried by the recoiling gun is equal in size and opposite in direction to that of the bullet. Because the mass of the bullet is so much smaller than that of the gun, the *speed* of recoil is much less than the speed of the bullet. In later chapters we shall see that many subatomic particles eject 'bullets' at very high speed.

Another example is provided by the rocket. This machine expels gases backwards at high speed, and it conserves momentum by propelling the payload forward. In deep space, away from gravitational forces, if the rocket is switched off, the payload will continue with undiminished momentum, and so the spacecraft will simply fly on with constant velocity (see also Fig. 1.2).

Another type of momentum concerns the rotation of bodies, and is called *angular* momentum. A heavy, rapidly spinning body carries a lot of angular momentum. The exact mathematical definition need not concern us, but an example is helpful. The angular momentum of a ring rotating about an axis through its centre, perpendicular to its plane, is given by the formula mvr, or mass × velocity × radius. From this formula one can see that the angular momentum is increased by making the ring heavier, increasing its radius, r, or spinning it faster. These qualitative remarks are generally true for spinning bodies of less regular shape.

Newton's laws tell us that, in the absence of external forces, angular momentum is also conserved. A good example of this is the spinning ice

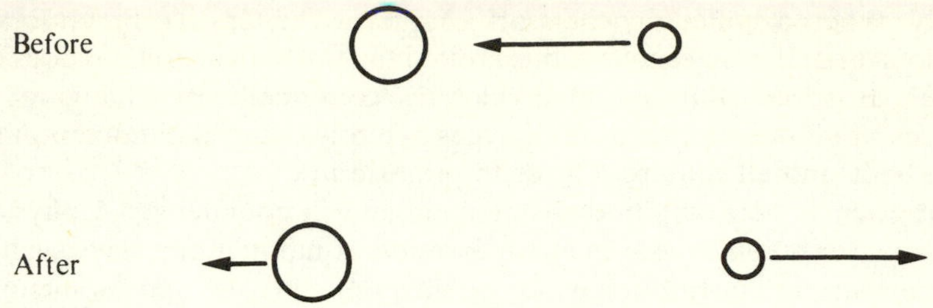

Fig. 1.2. Conservation of momentum. The forward momentum of the light ball is reversed when it bounces off the heavy ball. Momentum is a vector, so a change in its *direction* constitutes an acceleration. Newton's second law attributes the 'reversing' acceleration to the force of impact. The same law applied to the *total* system says that the *total* momentum remains unchanged (friction neglected). Thus, the change in momentum by the light ball is exactly compensated by an oppositely directed momentum change in the heavy ball, which recoils.

skater who, by contracting the arms and therefore concentrating the mass closer to the axis of rotation, conserves angular momentum by increasing the rotation rate. If this procedure was adopted for the rotating ring, it would mean that shrinking the radius r would increase the rotational velocity v by the precise amount necessary for the product mvr to remain unchanged, e.g. halving the radius doubles the rotation rate.

Angular momentum is a vector, and is described by drawing an arrow along the rotation axis. There is a convention about the direction of the arrow, best remembered by imagining the rotation of the hands of a clock. The arrow describing the hands' angular momentum points *into* the front of the clock (see Fig. 1.3).

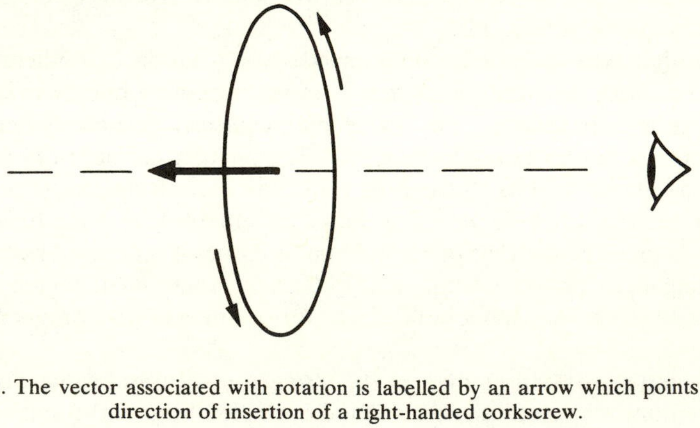

Fig. 1.3. The vector associated with rotation is labelled by an arrow which points in the direction of insertion of a right-handed corkscrew.

The final conservation law which must be remembered is the conservation of *energy*. Energy is present in the world in many forms; for example as motion, heat, electricity, or stored up behind dams or in pieces of stretched elastic. Energy can be readily converted from one form to another, but never created or destroyed, so long as no external forces act. As a random example consider an arrow shot from a bow. The archer supplies energy from his body to stretch the bow string, where it is temporarily stored as elastic energy. When the string is released the energy is rapidly converted into the motion of the arrow. When the arrow strikes the target, the motion is arrested and the energy converted into heat, sound and the work done parting the material of the target. At all stages of the sequence, the total energy in the system is the same.

Another example concerns a heavy body lifted from the ground and then dropped. The work done during the lifting process converts to potential energy stored invisibly in the body by virtue of its elevated location. It can easily be released again by dropping the body, whereupon the potential energy steadily converts into energy of motion as the falling body picks up speed and finally dissipates all this energy as heat, sound, etc. when it strikes the ground. At no stage is energy gained or lost as far as the *total* system is concerned.

Physicists believe that the three conservation laws – momentum, angular momentum and energy – suitably generalized, are absolutely obeyed in all natural processes without exception. This is a very strong statement, and one that plays a vital role in ordering the behaviour of subatomic matter. Other quantities are also found to be conserved in many processes (some of these will be discussed in Chapters 5 and 6) and some in all processes studied so far, but none has quite the entrenched status of the three dynamical quantities described here. As we shall see, their conservation is deeply connected with the structure of space and time, which places them in a class apart from other quantities which just seem to happen to be conserved for no special reason.

Armed with the vital notions of vectors, forces, momentum, angular momentum and energy, we shall now begin an examination of the nature of different forces, starting with the most familiar.

1.2 *Gravity*

It is frequently said that gravity is the most mysterious of all forces, and yet it was the first to be systematically studied and understood at a level of mathematical rigour which remained unchallenged for two centuries. The aura of mystery surrounding gravity has several origins. First, gravity controls the motions of the planets and stars, which imbues the subject with a sort of celestial reverence. Secondly, its extreme weakness is in marked contrast to 'everyday' forces, such as those of an electrical nature. If it were not for our proximity to the vast bulk of the planet Earth, we should not notice gravity at all. Because of this feebleness our technology is not able to harness or manipulate gravitational forces as it does others: it remains untamed. The weakness is all the more paradoxical when it is found that there might exist in the universe objects, such as collapsed stars, in which gravity overwhelms all else. A third reason for mystery emerges from the very nature of the currently accepted theory of gravity, which treats a gravitational field as a geometrical distortion of space–time. This property sets gravity apart from other forces and opens up the possibility of bizarre physical effects involving space and time, such as those associated with black holes.

Although different from other forces, gravity is not really mysterious in the sense of the unknown. Mathematical complexity has prevented a full study of all but a limited range of situations involving gravitating bodies, but the framework of the theory is available for those resourceful enough to exploit it.

Isaac Newton's great contribution was to deduce the nature of the force of gravity exerted by an isolated body such as the sun. He supposed that gravity acts between all material bodies in the universe with a force of attraction that diminishes with their separation in the celebrated inverse square law fashion. He assumed that the force was also proportional to the quantity of matter, or mass, contained in the body, and to the mass of the other body that experiences the force. Each body acts on the other with equal and oppositely directed attractive force. In mathematical symbols

$$F = \frac{GM_1M_2}{r^2} \tag{1.3}$$

or force of gravity = constant × mass of one body × mass of other body all divided by the $(\text{separation})^2$. The constant G is known as Newton's universal constant of gravitation, and it determines the *strength* of gravity between two given masses: gravity is weak because G is small (6.67×10^{-11} Nm^2 kg^{-2}). The force between two 1-kg masses placed 1 cm apart is less than one millionth of a Newton. The constant is called universal because as far as we can tell the gravity between two bodies of a given mass and separation is the same throughout the universe.

In spite of its extreme weakness, gravity is cumulative in power. We know of no objects whose gravity is repulsive, so that as more and more matter is aggregated together, so the mass terms in Equation (1.3) grow larger and the force F grows with them. When the bodies are of astronomical size, the force of gravity dominates over all others by sheer quantity of material. It is possible to measure the gravitational force exerted between two large, spherical masses even in the laboratory. This experiment confirms that the constant G is the same between metal spheres as between the Earth and moon.

Although it controls the motion of the universe and determines the structure of galaxies and stars, gravity does not seem to play an important role in the microscopic structure of ordinary matter itself, because of its extreme weakness. However, in very massive and compact objects, such as certain types of collapsed stars, the force of gravity may be strong enough to crush even atoms. Under these circumstances its effect on the microscopic behaviour of matter is still not fully understood.

1.3 *Magnetism and electricity*

Gravity is such a commonplace form of action-at-a-distance that its existence occasions little puzzlement or surprise. In contrast, the behaviour of magnets always seems rather novel and curious. Most people's initial experience with magnetism is the compass. If a magnetized needle is pivoted horizontally, and turned to some arbitrary angle, a mysterious force acts on it to rotate it until it aligns roughly north-south. The force is apparently produced from deep in the Earth and in some way reaches up from beneath the ground to grab at the magnet.

Magnetic action-at-a-distance can be graphically illustrated by holding two metal bar magnets close together as shown in Fig. 1.4. It is always something of a surprise to feel the two bars pushing each other away, or pulling together if their relative orientation is reversed. Because of the association with the compass we still call one end of the magnet the north pole and the other end the south pole. Simple experimentation reveals that like poles repel each other while unlike poles attract. More careful experiment and measurement shows that the *strength* of the force varies with the distance between the two poles in exactly the same way as the force of gravity between two small masses; that is, an inverse square law. However, a curious and unexplained feature about magnetic poles is that they never seem to occur singly, but always in couples – one north, one south. This means that the force between two magnets is really the combination of *four* forces between each of the four poles. Because two are attractive and two are repulsive, they tend to cancel each other, so that the total force actually falls off more rapidly than the square of the distance.

Although commercial magnets are usually made of iron, most materials are found to have some magnetic action, however slight. In some cases, such as iron, the magnetism can readily be imposed on the metal

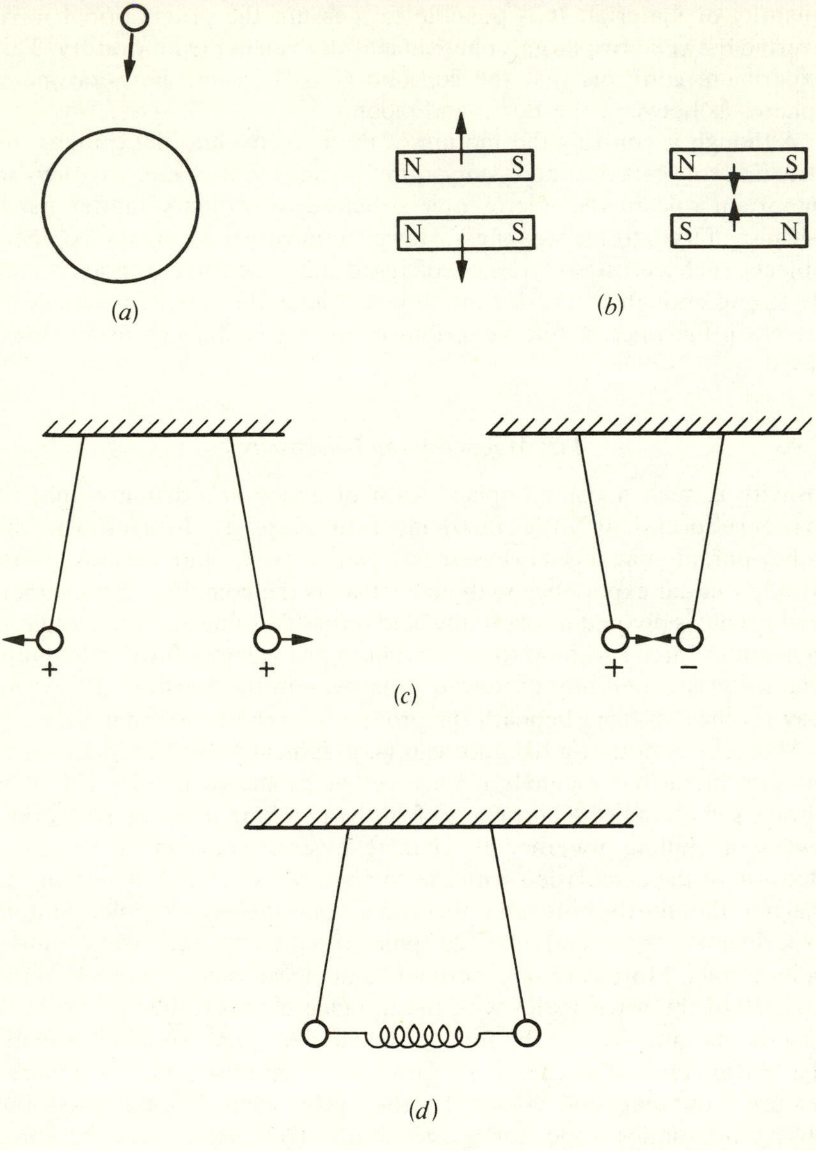

Fig. 1.4. Action-at-a-distance. (*a*) Gravity reaches out across empty space to pull on all other matter with a force that diminishes with distance in a mathematically precise fashion (inverse square law). (*b*) Magnetism both attracts and repels other magnetic substances even though they are not in physical contact. Magnetic poles come in two types, north and south. Their mutual force also diminishes with separation. (*c*) Electricity provides a third example of action-at-a-distance. The like charges on the suspended balls forces them apart, and the unlike charges draw them together, with a force that yet again diminishes with separation in an inverse square law. (*d*) A similar effect can be produced by inserting a spring between the balls, though in this case the cause of their repulsion is visible.

using another magnet, and it usually remains on the iron permanently. More often, however, the magnetism only appears when a permanent magnet is nearby, and disappears again when it is removed. In spite of these differences, the ubiquity of magnetic action in materials suggests that magnetism is a rather fundamental property of matter, to whose basic constituents we should look for an explanation.

Coming now to electricity, modern domestic appliances have made it seem as commonplace as tap water, yet its significance ranges far beyond television sets and cookers; electricity lies buried in the deepest confines of matter. The power of electricity is evident to all through the electric motor and the thunderbolt. The simple expedient of combing one's hair can produce an electrical force which rivals the gravity of the entire Earth, making the individual hairs stand upwards in defiance of their own weight.

In daily life the forces exerted by electricity are usually through electric currents and their *magnetic* effects, with which electricity is inextricably interwoven, as we shall see in the next section. Pure electric forces are not especially familiar outside the laboratory, except for party tricks such as sticking rubber balloons to ceilings. Early investigators soon discovered that, as with magnetism, there are two types of electricity, arbitrarily called positive and negative. Positive and negative electricity are analogous to north and south magnetism, but unlike the latter, it is possible to isolate positive from negative. Both seem to be invisible, but ordinary objects can be 'charged' with either of these mysterious 'substances', the presence of which manifests itself through numerous phenomena such as luminous clouds, painful shocks or sudden sparks.

Electricity can apparently come and go but a quantity of positive electricity is always created or destroyed along with an equal quantity of negative electricity, so the total numerical amount (quantity of positive − quantity of negative) never changes. This is sometimes expressed by saying that electricity is *conserved*. It can be produced easily by friction, for example rubbing ordinary articles with fur, and appears spontaneously in huge quantities in the atmosphere to power electric storms. It travels easily through some substances, for example metals, called conductors, but is blocked effectively by others, such as rubber, called insulators. The early investigators therefore collected electricity, by a variety of ingenious devices, on metal objects (typically a ball of copper) separated from their surroundings by insulating material. This enabled them to accumulate large quantities of electricity in one place to study its effects, without it running away into the ground, which it tends to do.

A body containing electricity is *charged*: it can be positively or negatively charged. If negative charge is added to a positively charged body, or vice versa, the total charge is reduced, because negative and positive cancel each other as with negative and positive numbers. If charge is added to a metal object through which it can move easily it seems to spread out rapidly across the surface, suggesting that it has a natural repulsion of itself and tries to dilute itself by migrating to any extremities to which it has easy access. This repulsive force can be directly observed by placing two metal balls charged with electricity of the same sign close to one another (see Fig. 1.4). It is found that a type of repulsive gravity operates to try and push the balls apart. If the charges are of opposite sign the force is one of attraction, as with gravity. Thus, electricity exerts an action-at-a-distance too. Moreover, the strength of the force between two small charged bodies varies with distance in exactly the same way as Newton's gravity: an inverse square law. The role of gravitational mass is played by the electric charge – double the charge and you double the force. We may write down an equation directly analogous to (1.3)

$$F = \frac{Ke_1e_2}{r^2} \tag{1.4}$$

where e_1 and e_2 are the magnitudes of the charges, r is their separation and K is a number which determines the strength of the force. K plays the role of Newton's constant of gravitation, G, and like that quantity, must be found by experiment. In section 2.8 we shall consider the actual numbers for K, e_1 and e_2 and compare the strength of electricity with that of gravity. Notice that if e_1 and e_2 are of opposite sign (one +, one −) the force F is *negative* which indicates attraction, but if they are of the same sign (both + or both −) then it is positive, indicating repulsion.

One fundamental question that must be asked is where electricity comes from. When it is produced by rubbing objects together, it is found that one object becomes positively charged and the other negatively charged. This suggests that both types of electricity naturally reside in the materials that are rubbed, and the rubbing process separates them. In fact, electricity resides in all ordinary matter, but usually in equal quantities of positive and negative charge, so that the net electrical effect is zero and we do not notice it.

The fact that even electrically neutral objects still contain charge, in equal quantities of + and −, can be demonstrated by bringing a negatively charged body close to an uncharged metal bar. The proximity of the negative charge repels the negative charge inside the neutral metal, and because electricity moves easily through metals, it retreats to the remote end of the bar, leaving the near end with a net positive charge. Thus, the electrically neutral bar has become *polarized*, with the total quantity of its positive and negative charge unchanged, but partly separated.

Electric and magnetic forces were extensively studied in the early part of the nineteenth century and investigators sought a more physical explanation of how it is that a force can act across empty space between two magnets or two electric charges. The basic problem is illustrated by Fig. 1.4*d*. If two metal balls are suspended side by side they will hang vertically, unaware of each other's existence. Suppose a compressed spring is now introduced between the balls and allowed to expand. The system will take up a new configuration with the balls hanging obliquely. The cause of the disturbance is visible: the spring is in contact with each ball, and provides a continuous communication between them through its coils. The force of the spring pushes the balls apart. Now suppose that instead of using a spring the balls are given some electric charge of like sign. The effect is the same, the balls repel each other and hang at an angle to the vertical, only in this case there is no visible means of support. What agency is responsible for communicating the electric interaction?

This question led to the discovery of a profound new physical concept, the usefulness of which has extended far beyond considerations of electricity and magnetism. It was introduced originally by Michael Faraday (British, 1791–1867) and extensively developed by James Clerk Maxwell (British, 1831–79). The model of the interaction which Faraday proposed was that electric charge is surrounded by an invisible electric *field*, the action of which is manifested when another charged body is brought nearby. Instead of regarding the electric action as reaching across the gap between two charged bodies, it is to be envisaged as a *local* interaction between charge and field. A charged body feels a force because it is in contact with the surrounding field. In this way, action-at-a-distance is converted into action by contact at two separated places: each charged body feels the field of the other as a local force.

Magnets can also be regarded this way, with each magnetic pole producing a surrounding magnetic field which acts on other poles to produce a magnetic force. In both the electric and magnetic cases we can think of the charge or pole as the *source* of the field. Near to its source the field is strong, but it weakens with distance as though it is being diluted. The corresponding force which the field produces also diminishes with distance from the source.

Of course, at this level, the introduction of the field is little more than a linguistic convenience, because it does not tell us anything new about the interaction between charges or magnets. The actual observable result – the force on the neighbouring body – is the same whether the field or action-at-a-distance concepts are invoked. In the next section we shall see that the field is much more than an alternative description of action-at-a-distance, and can lead to new and exciting phenomena. For the moment, however, it is instructive to develop the language and ideas of fields in some detail.

There is a sense in which the field has a *shape*, and drawings of fields associated with various source arrangements are often given in text-books. To map the shape of an electric field, one measures the force acting on a small test charge (i.e. one small enough so that its own field will not greatly disturb that of the charge distribution which we are trying to investigate) at a range of points in the region of interest. Fig. 1.5 shows the characteristic patterns associated with some simple types of charge arrangements. The essential feature is the presence of a

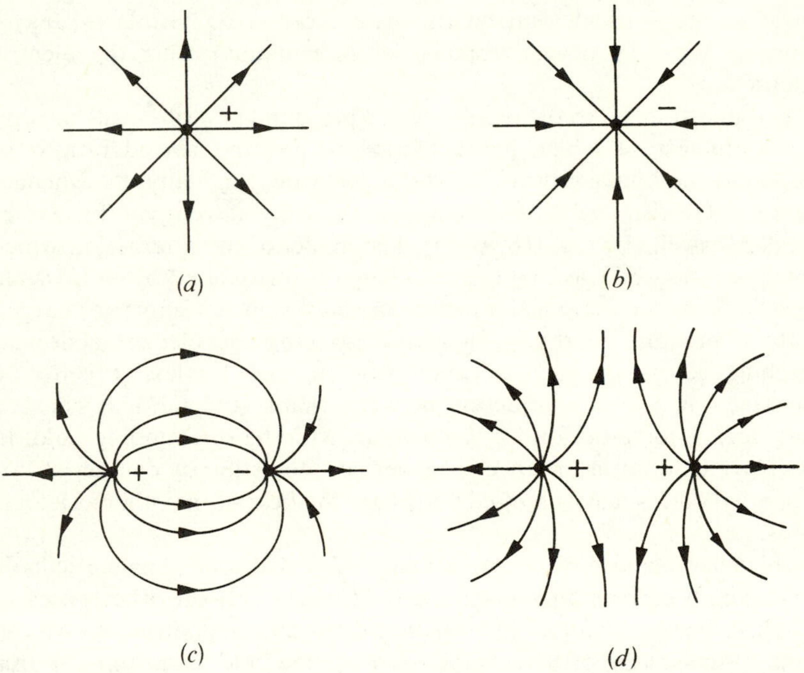

Fig. 1.5. Field shapes. The electric field surrounding various charge arrangements can be represented by a pattern of lines showing the direction of the force acting on a small, positively charged test particle placed at that point. A single isolated charge acts as a source (*a*) or sink (*b*) of a symmetrically shaped field. The field of two charges has a more complicated shape – each charge disturbs the field of the other. In (*c*) the unlike charges attract, and this is reflected in the many lines of force which link them together, like a network of stretched elastic. In (*d*) like charges repel; no lines connect them. Instead the lines turn away from each other like a cushion of compressed elastic between the charges.

systematic pattern of lines (sometimes referred to as lines of force) radiating out from the sources. Lines can only begin and end where charges are present: the field cannot abruptly disappear in empty space. In fact the field extends away from the source towards infinity, and strictly speaking there is no place where a sufficiently sensitive measurement would not detect it. This means that there is no natural length scale associated with the field beyond which we can say the intensity is small. Limitations of smallness are entirely a consequence of the measuring apparatus. This point about the absence of an inbuilt scale is an important one which will turn out to play a vital part in distinguishing electric (and magnetic) forces from others.

The arrows on the field lines show the direction of the force. By convention the test charge is taken to be positive so the arrows point away from positive charges (which repel the test charges) and towards negative ones. One crucial feature is that the electric field is a field of *force*, so that it has a direction associated with it. It is a *vector* field. Contrast this with a field of, for example, heat, where there is a temperature distribution in the vicinity of some heat source that can be measured by placing test thermometers near it. There would be no lines in such a map, only numbers attached to each point.

Similar considerations apply to magnetic fields, whose shapes may be mapped by small test magnets. The magnetic field is also a vector field of unlimited range, though, as already remarked, because only two-pole fields have so far been found in nature the field weakens faster with distance than the electric field of an isolated charge. The field lines can actually be observed rather spectacularly by surrounding a magnet with iron filings. Each small piece of iron becomes temporarily magnetized by the field in the particular direction of the field at its location, and acts as a tiny test magnet. All these little magnets then stick to their neighbours and organize themselves collectively along the force lines.

The field shapes shown in Fig. 1.5 are for sources placed in otherwise empty space. If materials are present these can distort the shape of the field dramatically. Crudely speaking, the field likes some materials and dislikes others, and the field lines tend to crowd into the former and avoid the latter. Electric fields have a particular predilection for metals such as copper, but dislike substances such as glass. Magnetic fields like iron, but dislike brass. It is possible to shield a region from the field by interceding material of the former kind. For example, there can be no electric field inside a closed metallic conductor such as a sphere of copper – the field lines crowd into the material of the sphere so do not penetrate into the interior.

If a lot of electric charge is placed at one end of a thin copper wire, the electric field lines crowd along the interior of the wire, so the force acting on any electricity in the wire is intense and tends to push it along the line of the wire, whatever shape the wire might be. When this happens one obtains an electric *current*. Until now we have only considered the fields due to static electric charge and magnets. When these entities move a whole new range of phenomena is revealed. This will be the subject of the next section.

1.4 *Electromagnetism*

In the 1820s, experiments were conducted by André Marie Ampère (French, 1775–1836), Jean Baptiste Biot (French, 1774–1862), Hans Christian Oersted (Danish, 1777–1851) and Felix Savart (French, 1791–1841) on the forces which act between *moving* electric charges, i.e. electric currents. It was found that an electric current produces a *magnetic* field. The shape of the field round a single straight wire is shown in Fig. 1.6; notice that it is everywhere *perpendicular* to the wire.

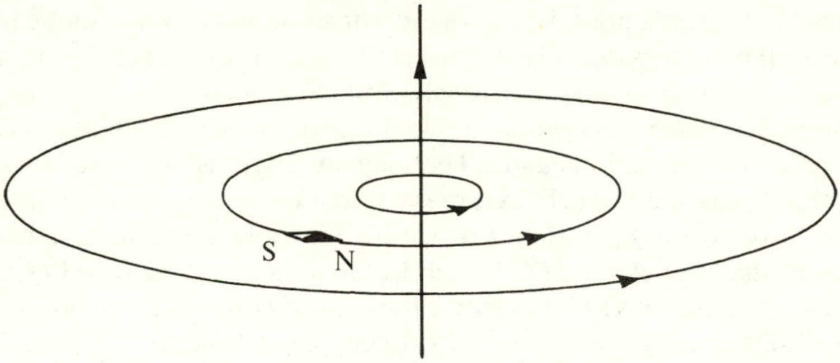

Fig. 1.6. Magnetic field due to electric current. At every point the field is perpendicular to the wire (see test magnet).

The field around a coil of wire (Fig. 1.7) is more interesting – it has exactly the same shape as that around a bar magnet. Such a field shape would also be produced by a single electric charge orbiting around in a circle.

Both these examples demonstrate a significant feature about the lines of magnetic force produced by moving electricity: they have no beginning or end. In the first example they are circles, in the second they are a sort of flattened ellipse shape, i.e. closed curves. Because lines of force cannot just end in empty space, they must either join up with themselves in closed curves, or end on magnetic sources (i.e. magnetic 'charges'). Magnetic fields associated with electric currents *do not have magnetic sources*. Look at the field around the bar magnet. There the lines of

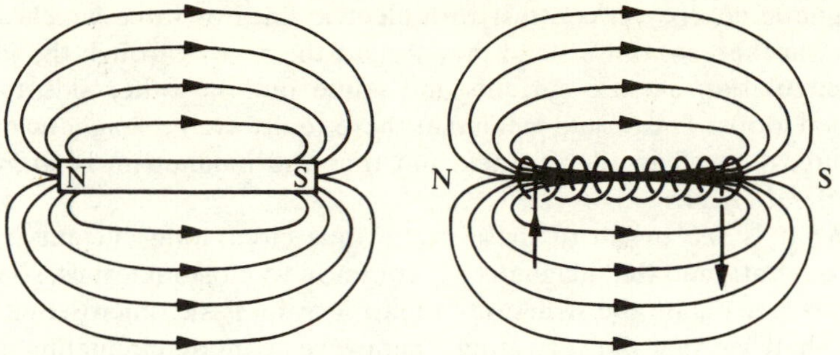

Fig. 1.7. Magnetic fields. The field around the current-carrying coil is identical to that outside the bar magnet. However, the force lines can be traced threading through the *interior* of the coil: they are closed loops. The absence of isolated magnetic poles suggests that if the lines could also be traced though the interior of the metal bar, they would be found to thread through tiny circulating currents on an atomic scale. The lines denote the direction of the force acting on a N pole test magnet.

force enter the material of the magnet near the north and south poles and seem to end there – the metal of the magnet seems to be a magnetic source. Yet if that were so, we ought to be able to cut a north pole out of the magnet and isolate it. If one does cut off the north pole, a new south pole instantly appears at the cut (see Fig. 1.8).

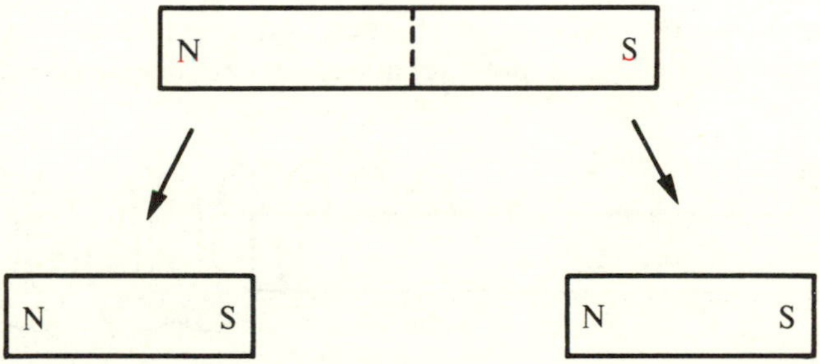

Fig. 1.8. Magnetic poles cannot be isolated. So far as we know all magnets occur in pairs (dipoles). Attempts to cut the N and S poles apart by sawing along the broken centre line only produce a new S and N pole, respectively at the end facing the cut.

One reason for this inseparability now suggests itself. If we ask the question, what happens to the lines of force when they enter the interior of the magnet, we might guess that they crowd up along the centre rather like the lines in the interior of the current-carrying coil. That is, the close similarity between the fields of the bar magnet and the coil suggests that the lines of force do *not* end inside the magnet on some magnetic charge (in contrast with electric lines of force which end on electric charges) but instead they thread their way through the interior of circulating electric currents and come out the other side to form closed loops. This would mean that there really are no magnetic charges (sources) at all in the magnet, and that the magnetism is *entirely an electrical effect*.

What is the origin of these mysterious circulating currents? It has been mentioned that magnetism is common to most materials, so we are led to investigate the structure of matter at its basic, microscopic level. We shall see that the very atoms themselves contain circulating electric charges. Magnetism on a macroscopic scale is in fact a gigantic, cooperative accumulation of billions of atomic magnets.

The reason why north poles are always apparently accompanied by south poles is because magnetism always seems to be produced by electricity. No magnetic charges have been discovered, although electric charges are everywhere. The reason for this asymmetry has long intrigued physicists. The asymmetry is still more baffling when the close symmetry between electric and magnetic phenomena is realized. For example, not only do moving electric charges produce magnetic fields, but moving magnets produce electric fields. If a magnet is jiggled around near a loop of wire it is found that an electric current flows in the wire, indicating that an electric field has been established along the wire. This phenomenon is known as induction, and was discovered by Michael Faraday (see Fig. 1.9).

In summary, moving electric charges can exert forces on magnets across empty space, and moving magnets can also exert forces on

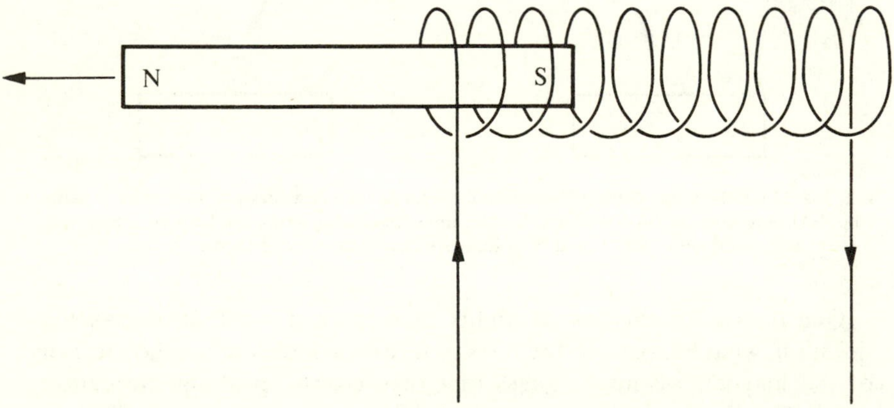

Fig. 1.9. Electromagnetic induction. If the magnet is suddenly withdrawn from the interior of the coil, the change in magnetic field threading the wire causes an electric field which drives a current round the coil. This principle is used to generate domestic electricity.

electric charges, causing them to flow along wires. Evidently electric and magnetic forces are closely interwoven, which suggests that a *unified* theory, involving just a single *electromagnetic* field, can be formulated. The decisive step in this direction was taken by Maxwell in the 1860s.

To appreciate this advance, yet another feature of the interplay between electricity and magnetism must be explained. As there are no magnetic charges, the effect of induction is not really caused by moving magnets so much as by a changing magnetic *field* (remember that the magnetism is really caused by electric effects). If magnetic charges did exist, and were made to flow along in a magnetic current, then this too would produce an electric field. On the other hand, in the experiments of Ampère and others, it was an electric current – a real motion of electric charge – that produced the magnetic field.

What Maxwell set out to do was to write down a set of mathematical equations which link together the strengths of the electric and magnetic fields with the strengths and arrangements of their sources, i.e. of the electric currents and changing magnetic fields. One equation predicts the magnetic field produced by an electric current, another predicts the electric field produced by a changing magnetic field, yet another gives the electric field produced by electric charge and a final one states mathematically the condition that the magnetic lines of force must not end (there is no magnetic charge). The form of these four equations was based on the experiments and observations already described, and they obviously entangle electric and magnetic effects together. Maxwell's great contribution was to notice that there was a *mathematical inconsistency* among this set of equations: one of them had to be wrong. They could be corrected, he found, by adding another term to the first equation. Physically the extra term corresponds to a way of producing a magnetic field other than from an electric current.

The physical effect that had been overlooked is that a magnetic field can be produced by *a changing electric field*, just as an electric field can be produced by a changing magnetic field. As an example of this effect, suppose one has two parallel metal plates a short distance apart with some positive charge on one plate and negative charge on the other. The presence of the charges will produce an electric field in the gap between the plates, so if the charges are varied then the field will change with time. A compass placed between the plates will be deflected, showing the presence of a magnetic field there produced by the variations in the electric field (see Fig. 1.10).

Parallel metal plates are known as condensers and are frequently used in electrical circuitry. In the simple arrangment shown in Fig. 1.10 a condenser is connected to a source of electricity such as a car battery. Inside the battery, lots of electric charge is trapped. When the switch is

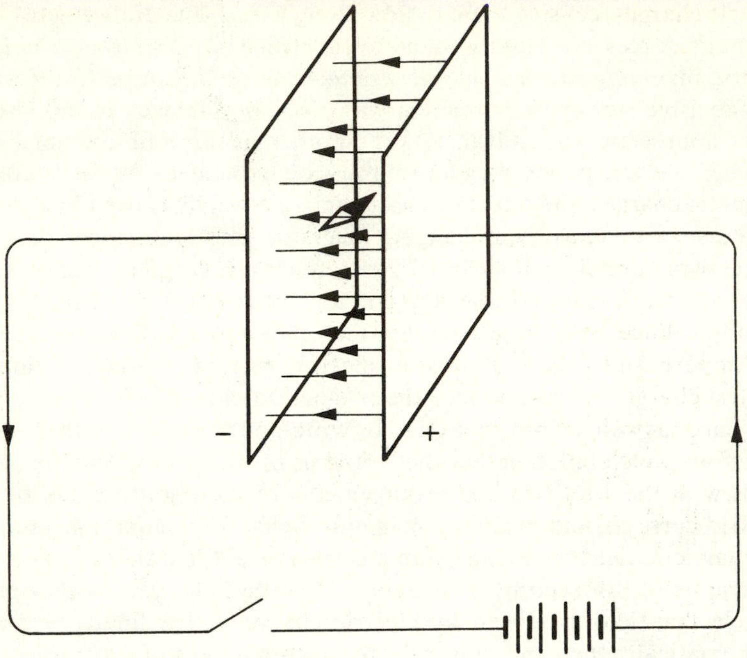

Fig. 1.10. When the switch is closed the electric battery delivers some charge to the plates of the condenser. As the charge flows into the plates along the wires so an electric field (arrows) suddenly builds up between the plates. Although no electric current flows across the gap, the *changing electric field* acts as a source of magnetic field, and this deflects the magnet. The way is now open for electromagnetic waves.

closed, the battery is connected via a wire to one plate of the condenser, so the pent-up charge bursts out under its own self-repulsion and is propelled by the electric field of all the rest of the charge down the wire to the plate. There it meets a dead end and stops. However, its influence still reaches across the gap in the form of its electric field, and this field repels the negative charge in the opposite plate and polarizes the other half of the circuit, driving the negative charge round into the positive terminal of the battery (to which it is also attracted by the positive charge there). After a short while the rearrangement is complete and the action ceases. Thus although no electric current can flow directly across the gap between the plates of the condenser, the current still effectively crosses it because of the operation of the electric field in the gap. The *changing* field therefore acts as an effective electric current, and the missing term in Maxwell's equations predicts that, just as with a real current, it too produces a magnetic field.

It is hard to overemphasize the significance of Maxwell's contribution.

First it demonstrates a remarkable symmetry between electric and magnetic fields – each acts as a source of the other – thus making the concept of a unified electromagnetic field more meaningful and elegant. Each point in space has two fields, electric and magnetic, but they are not independent, being conneced through Maxwell's equations. Thus they are better regarded as two *components* of a single, unified, *electromagnetic* field.

The second consequence of adding the missing term is profound. It was already known that a changing magnetic field acts as a source of electric field, but Maxwell showed that a changing electric field also generates a magnetic field. We therefore have a completely new and exciting possibility – that electric and magnetic fields can exist *without the need for electric currents to produce them*. The reason for this is that both the electric and magnetic fields *can feed off each other*, each acting as a source of the other. Such a possibility naturally involves only *changing* fields and has a sort of self-sustaining property. An electric field starts to increase and generates a magnetic field; the increase of this in turn generates another electric field which creates more magnetic field and so on. The precise way in which the changing fields interlock is encapsulated in Maxwell's equations, and they can be analysed mathematically in a standard way to see what will happen.

One very simple type of continually changing field which turns out to be a solution of Maxwell's equations has the form of a vibration or wave motion. In fact, the equations predict that an electromagnetic field can propagate as a wave disturbance through empty space, at a fixed speed which depends on the electric and magnetic properties of space. These properties can be measured in the laboratory and the speed estimated. It works out at about 300 000 km s^{-1}, which is very fast, but also very familiar. It is precisely the speed of light. Maxwell realized this and concluded that light is a form of electromagnetic wave.

Experiments do indeed show that light has wave properties (see Chapter 2) and it is possible to use these experiments to measure the length of the light waves (distance from 'crest to crest'). Typical optical wavelengths are about 10^{-7} m, which is small enough to prevent our noticing in daily life that light has a wave structure.

The wavelength of light has a certain range, which explains why light can have different *colours*. The familiar rainbow spectrum of colours is simply a spectrum of wavelengths, the red colour having a wavelength several times longer than the blue. Maxwell reasoned that electromagnetic waves could exist with all manner of wavelengths, even those that are much shorter or longer than visible light. Some 20 years later the German physicist Heinrich Rudolph Hertz (1857–94) managed to generate electromagnetic waves with wavelengths of the order of *metres*.

We now call these long wavelength electromagnetic waves *radio waves*. Since that time a whole spectrum of other wavelenths has been been discovered varying from the shortest wavelength gamma rays (less than 10^{-12} m) through X-rays and ultra-violet, visible, infra-red and microwave, to radio. Fig. 1.11 shows these different wavelengths schematically on a diagram. They are all examples of the same basic wave phenomenon.

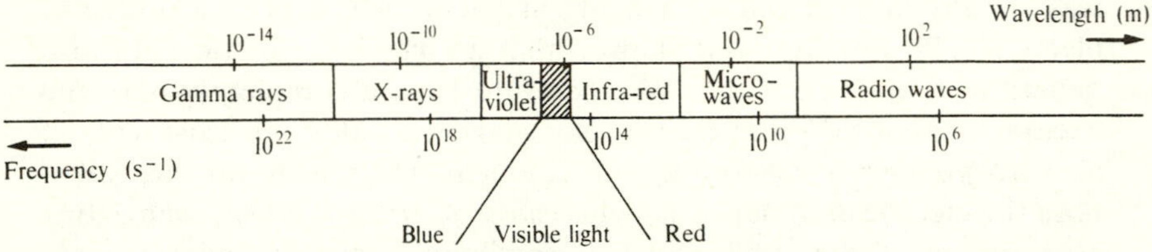

Fig. 1.11. Electromagnetic spectrum.

The importance of the discovery of self-sustaining electromagnetic fields such as wave disturbances is that it bestows upon the field an *independent existence*. Before the additional term was added by Maxwell, electric and magnetic fields were inevitably tied to their sources in a slavish way, so that they represented little more than a fancy description of how electric currents act upon each other at a distance. The fields themselves did not seem to have any properties that were not already present in the electric currents. After Maxwell the electromagnetic field became liberated and assumed the status of a physical system in its own right, existing alongside, and interacting with, electric charges and currents. This is a revolutionary new concept of nature that goes far beyond the unification of electricity and magnetism. The idea of a field as a separate entity may be hard to grasp, but is a vital ingredient in understanding the forces of nature.

Although self-sustaining once it is started, we have to comprehend how an electromagnetic wave is set off in the first place. Just as water waves or sound waves are produced by a disturbance of the water or air, so electromagnetic waves are produced when something disturbs the electromagnetic field; for example, a moving charge or changing electric current. The wavelength and frequency of a sound wave is determined by the frequency of vibration of the source (e.g. an oscillating guitar string produces a certain definite note). Similarly the wavelength of an electromagnetic wave is determined also by the frequency of the disturbance. In a radio aerial, an electric current oscillates up and down at typically 1 MHz (1 Hz, short for Hertz, is one cycle per second), producing radio waves of the same frequency.

Maxwell's equations tell us that in empty space, electromagnetic waves always travel with a fixed velocity – the velocity of light. When an electromagnetic wave passes a point in space, the electric and magnetic field at that point oscillates with a certain frequency. If the wave is a long one, it will take proportionately longer for the field to execute one cycle. The length of the wave is related in a simple way to the frequency: speed = frequency × wavelength. In symbols we write this

$$c = \nu\lambda. \tag{1.5}$$

Using this relation we may convert between wavelength and frequency. For example, BBC Radio 1 is broadcast on the 275-m wavelength, which is equal to $3 \times 10^8/275 = 1.09 \times 10^6$ Hz.

It is instructive to compare the pattern of an electromagnetic field around a system, such as a radio aerial, which is emitting radiation, with that surrounding a static electric charge, or a steady current. As explained on pp. 11 and 14 the static electric and magnetic fields obey the inverse square law of force. That is, the strength of the fields falls away as $1/r^2$, where r is the distance from the source. In contrast the electromagnetic wave (radiation) strength falls off much more slowly, as $1/r$. That is why electromagnetic signals can travel to great distances.

This leaves us with the question of the origin of light and the shorter wavelength waves, which correspond to frequencies in excess of 10^{15} Hz. Evidently at some fundamental level in matter there exist electric currents that can oscillate with these enormous frequencies. The nature of these currents will be described in the forthcoming chapters. There is also the fascinating question of whether all electromagnetic disturbances were necessarily started off by electric currents somewhere, or whether some of them have always been around. This raises profound issues about causality and the creation of the universe which are beyond the scope of this book.

It was long appreciated that electric and magnetic fields store *energy*. To create a strong electric field, by placing a lot of charge on a metal ball, for example, it is necessary to expend energy in forcing all the charge together on to the ball against its natural repulsion. This energy reappears in the electric field, and could if desired be released again by allowing the charge to flow away from itself along a wire or something, to where it could drive an electric motor. Thus energy can be transferred from electric charge and current into electric and magnetic fields, and vice versa. In an electromagnetic wave, this field energy is *transported* through space in the oscillating fields of the wave. For example, a radio transmitter pumps electrical energy into the aerial in the form of a current. Some of this energy is released and travels away in the wave, to be absorbed by electric charges in the receiving aerial. The transmitter loses energy and the receiver gains some of it. The amount of energy stored in an electromagnetic wave can be deduced from Maxwell's equations.

The same equations show that the wave carries *momentum*, and it can produce a recoil in any object with which it impacts. A good example of this effect is the tail of a comet, which consists of gas blown off from the head of the comet by the pressure of light from the sun (the comet tail always points away from the sun). Some engineers have suggested spacecraft designs in which a ship is equipped with a large 'sail' and is blown through space by the pressure of sunlight.

Clearly, the electromagnetic field behaves very much like any other mechanical system, having its own type of motion, transmitting, absorbing and releasing energy and interacting with matter. Sometimes it is hard to accept this because the field seems so insubstantial and tenuous, especially because we cannot see it. Nevertheless, the physics of the electromagnetic field is very similar to that of other continuous systems, such as fluids, and the enormous range of electromagnetic phenomena in both physics and engineering are thoroughly well understood in terms of Maxwell's brilliant theory. We owe not only radio, television and telecommunications in general to Maxwell's mathematics, but also an advance in fundamental physics of the first magnitude.

1.5 *Relativity*

In Section 1.3 it was pointed out that an electric current generates a magnetic field. The current itself is a flow of electric charge, and the analogy with the flow of a river comes to mind. When we discuss the current of a river, we normally mean the motion of the water past a fixed point on the bank. However, if we are in a boat going upstream, the effective current past the boat is greater than that on the bank because of the motion of the boat relative to the bank. On the other hand, if the boat is drifting without power, it will float downstream with the current, so the water will not flow past the hull at all. From the mechanical point of view what matters to the crew of the boat is not their motion relative to the bank but that relative to the water.

Similarly with an electric current, what matters is the *relative* velocity between the moving electric charge and the observer. What one observer regards as a stationary electric charge, another would regard as a current. In this case the second observer will see a magnetic field but the first observer will not (though both see an electric field). Thus, the nature of electric and magnetic fields are closely connected with the *frame of reference* from which they are observed. Maxwell's theory of electrodynamics provides the mathematical machinery needed to *trans-*

form the fields as viewed from one frame of reference to those observed when viewed from another.

Great importance is attached to the way in which physical theories transform between reference frames, and in the late nineteenth century much attention was directed to the study of the transformation properties of Maxwell's electromagnetism in this respect. It was discovered that the transformation properties are fundamentally different from those that arise in Newtonian mechanics. When this was realized it was thought that electromagnetic phenomena must violate the basic principle of the relativity of motion. This principle, known since the time of Newton, says that absolute uniform motion through space has no meaning; only *relative* to some other material body does it make sense. Thus, we measure the speed of a car relative to the road, the speed of a train relative to the track, and so on. These measurements do have meaning in spite of the fact that the Earth is in motion round the solar system, the sun moves around the galaxy and the galaxy moves away from other galaxies as the universe expands. In a world full of motion, only that motion which is gauged against other bodies is relevant. To say that the Earth, sun, galaxy, or anything else, is 'at rest in space' is totally devoid of physical meaning. Or so it was thought. Because Maxwell's fields did not transform in the same way as the properties of material bodies when viewed by two observers in relative motion, it seemed that when matter and radiation interact, they ought to do so in a way that really could reveal whether a body was at rest in space, or moving at a certain speed in a certain direction in an absolute sense.

Experiments to measure, for example, the speed of the Earth through space, met with persistent failure, not because the technology was inadequate, but because the answer always came out as zero, even though it was known that the Earth at least orbits around the sun, so cannot possibly be at rest always. Independently of the experiments the great German physicist Albert Einstein (1879–1955) reconsidered the whole subject of matter and motion, and their relation to electromagnetism. In 1905 he proposed that, above all, the relativity of motion should remain unassailable. As an alternative, Einstein preferred to change the laws of motion for material bodies – a bold step, seeing that they had been constructed by Newton himself. To make the transformation properties (between reference frames) of Maxwell's electromagnetism compatible with the laws of motion, Einstein found it necessary to modify the accepted structure of space and time.

One of the consequences of the new space and time concerns the behaviour of electromagnetic waves. Maxwell's theory predicts that they propagate through space with a definite speed determined by the electric and magnetic properties of free space. On the basis of his new theory of relativity, Einstein claimed that an observer would measure this particular speed for light, *quite irrespective of his own speed or that of the source of light*. Such a seemingly paradoxical assertion – that the speed of light is the same in all reference frames – is totally at variance with everyday experience involving, say, sound waves. Nevertheless, it is found to be correct by many experiments. Among other things, this curious behaviour of electromagnetic waves prevents any material body from ever travelling at, or exceeding, the speed of light (see Fig. 1.12).

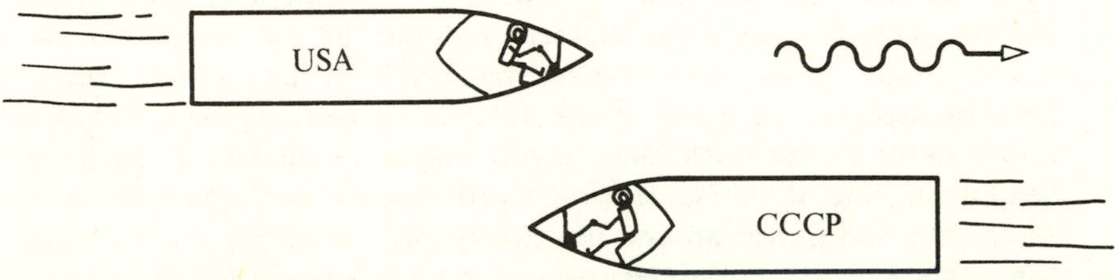

Fig. 1.12. Light speed is the same for everyone. The astronaut chases the light pulse. It gains on him by 300 000 km every second. The cosmonaut flees the *same* pulse. It leaves him at 300 000 km s^{-1}. The astronaut accelerates ('I'll overtake it yet!'), but the pulse continues to recede at 300 000 km s^{-1}. He is no better off than the cosmonaut driving hard in the opposite direction!

The inability to travel faster than light is not merely a mechanical inconvenience, but has a far deeper significance. For example, consider the electric field at some particular place due to the presence of an electric charge some distance away. Now suppose the electric charge is moved suddenly a little closer; the electric field will abruptly increase. One can ask the question, does the field change *immediately* the charge is moved (i.e. simultaneously)? Maxwell's equations show that this does not happen. Instead, the displacement of the charge causes a sudden disturbance of the field locally, and this disturbance spreads outwards at the speed of light. It follows that there is a definite *delay* before the strength of the electric field at a distant place responds to the relocation of the charge. This effect is called *retardation*. Picturesquely, it tells us that the information that the charge has been moved cannot get through until the field disturbance has *propagated* at a fixed speed to the distant place. It is found that no electromagnetic disturbance can travel faster than an electromagnetic wave (i.e. light). The picture of the field as an independent mechanical system therefore means that there can be no *instantaneous* action-at-a-distance between electric charges and currents. The action must be passed from one region of the field to the next at a definite speed – the speed of light.

In its full generality, Einstein's theory of relativity predicts that *no information or influence whatever* can exceed the speed of light. Put another way, electromagnetic waves play a central and fundamental role in the *causal* structure of space and time. Cause and effect can only be connected by slower-than-light signals or influences. The effect on the measurement of distances and time intervals due to this limitation can give rise to bizarre phenomena such as the dilation of clock rates and the apparent contraction of rigid rods. Furthermore, in this new arrangement, space and time become closely interwoven into an inseparable combination called space–time. Whereas in Newtonian theory space and time are quite distinct entities, in relativity theory they are on a more equal footing.

The unification of space and time into space–time brings with it a powerful new conceptual tool: the space–time map. Just as ordinary maps display a panoramic view of a whole patch of countryside, so a space–time map shows a complete region of space–time. It is useful for displaying the history of objects as they move about. Suppose we take a movie film of a moving object, separate the frames and stack them up on top of one another in chronological sequence. Looking down through the stack one will see a winding track – a record of the experiences of the object as it moves about. This track is known as the particle's *world line*. Points higher up the track record later events, so one can use the vertical distance as a measure of elapsed time (though the time told on a clock carried by the object itself will disagree with this because of the time dilation effect). The frames are horizontal slices through the stack representing space at one instant of observer (camera) time.

Fig. 1.13*a* shows the space–time map in which a runner sprints around a circular track, gradually slowing up from fatigue. In many cases the motions of interest will be confined to one dimension and we can draw the map as in Fig. 1.13*b*, which shows a light, fast moving ball striking a heavy stationary one and rebounding, causing a slight recoil (the situation also depicted in Fig. 1.2). As a final example, Fig. 1.13*c* shows schematically two motorists who drive to a certain distance apart, when one signals the other by radio (wavy line) after which they depart. Space–time maps will prove to be of great importance in later chapters for visualizing how subatomic particles interact with each other.

In everyday life, where speeds are so much less than that of light, the strange effects predicted by Einstein's theory are too small to be noticed. Why, then, is relativity important? We shall see that although human vehicles do not travel at anywhere near the speed of light, many subatomic particles do, and their behaviour can only be understood within the framework of the theory of relativity.

One result of the theory that is crucial for microphysics concerns the

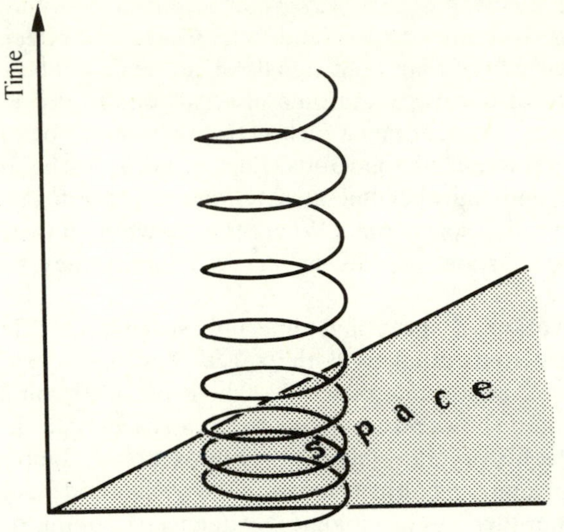

(*a*)

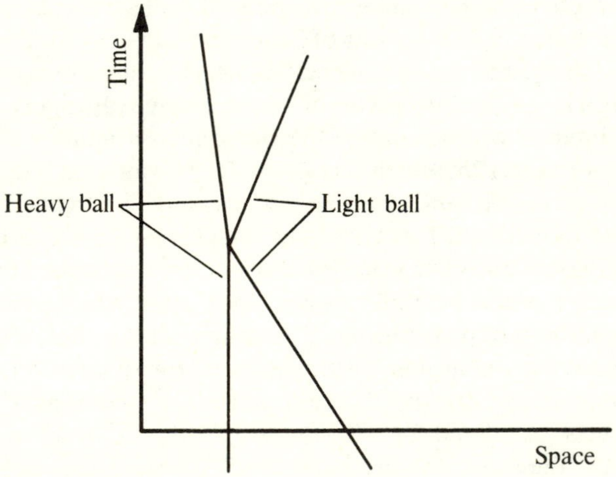

(*b*)

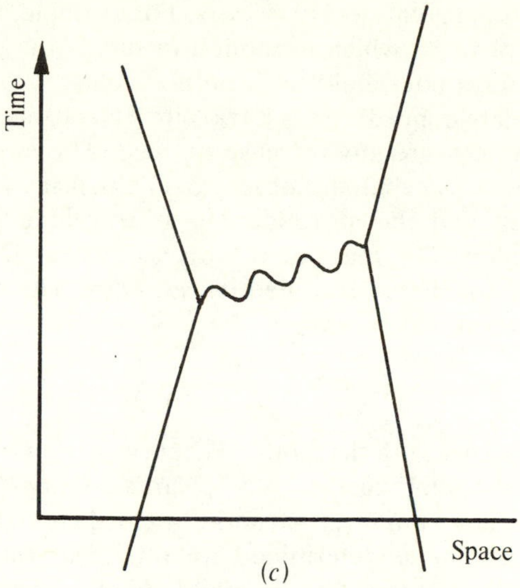

(*c*)

Fig. 1.13. Space–time maps. Horizontal slices give an instantaneous representation of space. The tracks on the map represent the histories of particles and waves. Viewing their itineries together in this way renders interactions readily visualizable. (*a*) The space–time path of a runner who gradually slows as he circles the track. (*b*) Map of one-dimensional motion, showing the collision process depicted in Fig. 1.3. (*c*) Radio communication (wavy line) between two observers who approach and then recede.

unification of the concepts of energy and mass. If a body is not permitted to travel faster than light it must protest in some way if we try to shove it harder and harder. What happens is that the *inertia* of the body grows with its relative speed. A particle moving close to the speed of light acquires a greatly enhanced mass, and this extra 'ponderousness' requires an ever greater expenditure of energy to overcome it. As the speed of light is approached, the mass rises without limit and acts as a physical barrier to further acceleration. Thus, in relativity theory, when a body is supplied with energy, the energy does not merely go into motion, it also gets converted to mass. The relation between energy delivered and mass acquired is a very simple one, contained in Einstein's most famous equation

$$E = mc^2 \tag{1.6}$$

or

$$\text{energy} = \text{mass} \times (\text{speed of light})^2.$$

Because the right-hand side of this equation contains the square of a very large number – the speed of light, c – it means that a small quantity of mass is 'worth' a great deal of energy. For example, 1 kg of mass can deliver 10^{17} J of work, which is enough to supply the power consumption of the average household for a million years.

Whatever relative speed a body is travelling at, an observer who travels with it will not measure any increase in mass. This is because uniform motion is relative, so that the observer might just as well regard the body as at rest and the surroundings as travelling very fast in the opposite direction. The mass of a body as measured by an observer travelling with it is therefore called its *rest mass*. The relation between total mass and rest mass is simply

$$m = \frac{m_{\text{rest}}}{\sqrt{(1 - v^2/c^2)}} \tag{1.7}$$

where v is the speed of the body. When $v = 0$, $m = m_{\text{rest}}$, but as v approaches c, the total mass m rises towards infinity. Notice, however, that should we have *both* $m_{\text{rest}} = 0$ and $v = c$, Equation (1.7) would read $m = 0/0$ which is an undetermined quantity, but can be finite. This means that a body *can* travel at the speed of light, c, provided it has zero rest mass. Its finite energy is then given by Equation (1.6). In later sections we shall see that particles with zero rest mass do exist and indeed, light itself can be considered as composed of particles of this sort.

An interesting relation follows if Equations (1.6) and (1.7) are combined together and squared

$$E^2 = \frac{m_{\text{rest}}^2 c^4}{1 - v^2/c^2} . \tag{1.8}$$

The right-hand side of Equation (1.8) can be rewritten

$$m_{\text{rest}}^2 c^4 + m^2 v^2 c^2 \tag{1.9}$$

as may be verified by simple substitution for m. Recall from p. 6 that in ordinary mechanics we call the product of mass and velocity the *momentum* of a body. Momentum plays an important role in relativity also, where it must be defined using the *total* mass m. Thus, denoting momentum by p we have $p = mv$. This enables expression (1.9) to be written as $m_{\text{rest}}^2 c^4 + p^2 c^2$, whence Equation (1.8) reduces to the curious result

$$E^2 = m_{\text{rest}}^2 c^4 + p^2 c^2 . \tag{1.10}$$

Further important information follows from examining the limiting cases of Equation (1.10). If the body is at rest, there is no momentum, so $p = 0$ and we have

$$E = m_{\text{rest}} c^2 . \tag{1.11}$$

This result tells us that even when a body is at rest it still contains an enormous quantity of energy. This energy is manifested as the *rest mass* of the body. Thus, rest mass may be regarded as trapped energy, concentrated into the body to supply it with inertia. Evidently, ordinary matter must be regarded as a reservoir of vast quantities of energy. In spite of man's knowledge about mass since time immemorial, it was not realized until Einstein's work that it represents a source of power of staggering proportions. On the Earth, we are sitting on a potential bomb equivalent to 10^{25} nuclear devices. In a later section it will be described how some of this trapped energy may actually be released.

Considering now the opposite limit in Equation (1.10) one sees that in the case of light itself (or other particles that move at the speed of light), we must put $m_{\text{rest}} = 0$ so that

$$E = pc \tag{1.12}$$

or energy = momentum × speed of light, a result which confirms that light carries momentum proportional to its energy and can therefore exert a pressure (see p. 26).

If a body is moving very fast, then its rest mass will be a negligible proportion of its total mass, and the p^2c^2 term will swamp the $m^2_{\text{rest}}\,c^4$ term in Equation (1.10). Under these circumstances it is a good approximation to neglect the rest mass altogether: the body's properties correspond closely to those of light, and we may use Equation (1.12). For example, at 99% of the speed of light, the term p^2c^2 is about 50 times larger than the rest mass term: that is, the body 'weighs' seven times its rest 'weight'.

The electromagnetic field plays a central role in the theory of relativity and hence in the structure of space and time. It is, however, only one of the forces of nature and consideration has to be given to gravitation. Newton's theory of action-at-a-distance is in direct contradiction with Einstein's relativity, because it enables a physical communication via gravity to be exerted instantaneously between bodies, rather than at light speed or less. Einstein reasoned that Newton's theory of gravity must also be incorrect and set about replacing it, a project that was not completed until another 10 years had elapsed. In 1915 he published the details of his general theory of relativity which extended the earlier theory to include gravity.

In many respects the general theory places gravity on a similar footing to electromagnetism. The gravitational field, although more complicated than its Maxwellian counterpart, has the same sort of transformation properties. Moreover, the new theory admits the existence of gravitational *waves* as a direct analogue of electromagnetic waves. In spite of strenuous efforts, none has yet been detected.* Waves of gravity are produced during turbulent or violent rearrangements of matter. They too travel at the speed of light. Thus, gravity cannot be used to achieve faster-than-light communication or influence.

In other respects gravity is, if anything, still more important for the structure of space and time than electromagnetism. The general theory of relativity in fact treats gravity *as* space–time, or more correctly as a manifestation of the geometry of space–time. Further consideration of the general theory is, however, beyond the scope of this book. The reader is referred to *Space and time in the modern universe* for a complete discussion of the general theory of relativity.

* As this book was going to press, indirect evidence for gravitational radiation emanating from a certain double star (known as a binary pulsar) was published.

2
The structure of matter

In the last chapter two familiar forces of nature that are well understood were briefly described. Gravity, which controls the motion of astronomical masses and serves to keep our feet on the ground, and electromagnetism which, it was claimed, is responsible for all the other familiar, everyday forces of nature. It may seem surprising that the tension of stretched elastic, the blow of a hammer, the push of the wind, the friction of a rough surface, the cohesion of a block of wood or metal, the pressure of a steam boiler, the explosive outburst of dynamite, as well as all the many other diverse examples of forces which surround us, are really electromagnetic in nature. Yet it is found to be so once an understanding of the structure of matter is obtained. Electricity and magnetism owe their origin to the microscopic constituents of matter, and we shall see that viewed on an atomic scale all the forces between atoms are essentially electromagnetic in origin.

2.1 *Atoms and molecules*

Most people are aware that all matter is composed of atoms and that these are exceedingly small entities. Words like 'atomic power' and 'atomic bomb' reinforce the impression that atoms are powerful and important, if somewhat mysterious, inhabitants of the microcosm.

The belief that simplicity lies at the root of complexity inspired a long, classical tradition that the enormous diversity of material objects around us ultimately consists of the accumulation of vast numbers of cohering, microscopic units of an identical, or at least similar, form. According to this idea, the large differences between substances as dissimilar as air, steel, glass and flesh is to be attributed solely to the arrangement of their constituent atoms (from the Greek *atomos* meaning 'indivisible'). As long ago as the fifth century B.C. the Greek philosopher Democritus (*c.* 460–357 B.C.) appears to have expressed the view that matter is composed of countless indestructible particles of various sizes, shapes and weights, engaging in ceaseless motion.

Much later, in the first century B.C. the Roman poet and philosopher

Lucretius (Titus Lucretius Carus, 98–55 B.C.) wrote in his poem *De rerum natura* that the atoms were never annihilated, but were hard, solid objects, individually devoid of properties such as colour, heat, taste and smell. They move about at great speed, collide and perhaps stick together in combinations according to their shapes. Naturally none of these early speculations was based on any sound evidence from experiment or observation, but the accuracy of the descriptions is astonishing, nonetheless.

Some early scientific evidence for the existence of a fundamental atomic unit came from careful experiment and measurement in chemistry. In the early nineteenth century it was found by John Dalton (British, 1766–1844) and others that during chemical reactions the weights of the substances which combine together are in certain fixed numerical proportions, suggesting that identical whole atoms of one substance stick together with whole atoms of another in definite small groupings. For example, carbon will combine with oxygen in two quite different ways. In one case $1\frac{1}{3}$ kg of oxygen combine with 1 kg of carbon to form a gas. In the other $2\frac{2}{3}$ kg of oxygen, i.e. precisely twice as much, combine with 1 kg of carbon to form another gas. We now know that the former gas is carbon monoxide, consisting of the union of one atom each of carbon and oxygen, while the latter, carbon dioxide, has two oxygen atoms glued to one of carbon. Hence the need for twice the weight of oxygen. The fact that in the monoxide $1\frac{1}{3}$ kg are needed for each kilogram of carbon reflects the fact that the oxygen atom is 4/3 times heavier than the carbon atom.

The union of two or more atoms in a chemical bond is called a *molecule*, and most substances encountered in daily life are molecular rather than atomic. Sometimes, in highly developed living matter, several thousand atoms are joined together to form long molecular chains of great complexity.

Further evidence for the atomic or molecular nature of matter comes from the behaviour of gases. In 1808, the French chemist Louis Joseph Gay-Lussac (1778–1850) demonstrated that when gases react together chemically, the *volumes* of each constituent gas required are also in simple numerical ratios (provided these volumes are measured at the same temperature and pressure). For example, 1 litre of oxygen requires precisely 2 litres of hydrogen to form 2 litres of water vapour, if no hydrogen or oxygen residue is to remain.

These facts are beautifully explained using the concept of atoms and molecules if the assumption, due originally (1811) to the Italian chemist and physicist Count Amadeo Avogadro (1776–1856), is made that a given volume of any gas (at a fixed temperature and pressure) contains the same number of molecules. This assumption is less surprising when it is realized that in a gas the molecules themselves are so small compared to their spacing that they occupy a negligible fraction of the total volume of the gas. Thus, when the atoms of two gases combine, the larger molecules thereby formed do not use up significantly more space. With this assumption it is easy to explain, for example, the oxygen–hydrogen ratio discussed above. Oxygen molecules consist of two oxygen atoms combined (O_2) and hydrogen is also diatomic (H_2). However, water consists of one oxygen atom stuck to two hydrogen atoms (H_2O). Thus each molecule of oxygen (two atoms) requires two molecules of hydrogen (four atoms) to make two molecules of water. According to Avogadro, this implies a 1 : 2 volume ratio, exactly as observed.

Later in the nineteenth century, chemists and physicists began to calculate what properties a gas could be expected to possess if it really did consist of billions of molecules rushing around chaotically. Clearly, these tiny particles will each carry energy and momentum, and when they collide with the walls of a container they will cause a small recoil. The cumulative, averaged effect of these countless tiny bombardments is to produce a steady force, or pressure, against the surface. Heating the gas speeds up the molecular motions and increases the force of the bombardment, which explains the well-known fact that the pressure of a gas rises when heat is added (see Fig. 2.1). The temperature of the gas is a measure of the molecular speed, so this increases also with the addition of heat. Furthermore, if a gas is introduced at high pressure into a container which is not rigid (such as a balloon), the molecular impacts will push the container walls outwards, and the work done in this activity is paid for by the gas losing some heat energy. This explains why a gas cools as it expands.

It is even possible to witness the molecular bombardment at work if the career of a microscopic particle, such as a pollen grain immersed in a liquid, is followed using a microscope. Some pollen grains are so small that random fluctuations in the motions of the fluid molecules cause momentarily uneven bombardment of the grains' surfaces and send the particles zig-zagging about. This phenomenon was first investigated in detail by the British botanist Robert Brown (1773–1858) and explained mathematically in 1905 by Einstein.

These experimental and theoretical facts lead to a convincing picture of gases composed of vast numbers of tiny particles in continual rapid and agitated motion, just as Democritus suggested more than two millennia ago. Of course, the evidence does not rest on these qualitative ideas. Detailed calculations by Maxwell, Rudolph Julius Emmanuel Clausius (German, 1822–88) and Ludwig Boltzmann (Austrian, 1844–1906) established the molecular theory of gases as a model for

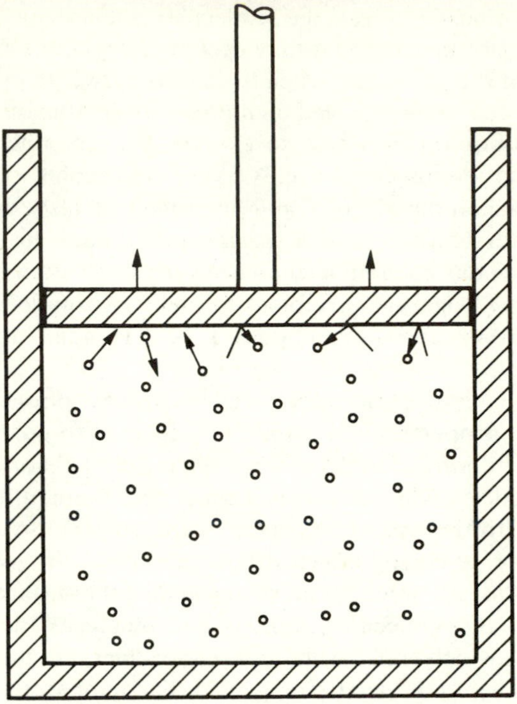

Fig. 2.1. Atomic theory of heat. Good evidence for the atomic nature of gases comes from calculations of the effect of their impact against a solid wall. As heat is added, the molecular motions become more vigorous and the pressure of the collisions against the underside of the piston is greater. The additional force raises the piston, releasing some of the heat energy in the process.

real gases capable of giving precise mathematical relationships between pressure, temperature, and so on. For example, by ignoring the size of the gas molecules and treating them as hard, point-like particles, all moving randomly and independently, with no interaction between them, it is straightforward to show that the average pressure of bombardment P is related to the gas temperature T and volume V through the simple formula

$$PV \propto T. \qquad (2.1)$$

The relation that the pressure of a gas (at fixed temperature) varies inversely with its volume had been known since the work of Robert Boyle (English natural philosopher, 1627–91).

Although the actual existence of atoms is no longer in doubt, many people are puzzled how physicists and chemists can obtain concrete information about them. Being so small, it is not possible to see or touch one, so that what we know about atoms comes mainly from indirect experiment involving some form of deduction. In fact, these days there are methods of detecting individual atoms, or at least processes which happen to them. One very useful way of examining small objects is to aim waves of some sort at them and watch for a disturbance in the wave pattern. To detect properly an object of a certain size, using this technique, it is necessary to employ waves with a wavelength smaller than the object concerned. This restriction is familiar in daily life: for example, short waves, such as ripples on a pond, would be greatly disturbed by the presence of a 1 cm thick post projecting through the surface, but such an obstacle would have little effect on the progress of an ocean swell.

In Chapter 1 it was explained how electromagnetic waves of all wavelengths could be produced by disturbing electric charges. Short waves, being of high frequency, require a correspondingly high rate of disturbance in the charges to generate them. To make electromagnetic waves as short as atoms it is necessary to disturb electric charges extremely violently, and a somewhat random, though practical means of doing this is to smash a highly energetic electric current into a fixed obstacle. The sudden deceleration of the charges causes a burst of short-wavelength radiation, such as X-rays. The waves are so short that it is possible to bounce pulses of these X-rays off individual atoms and detect them as spots on a photographic plate.

The use of X-rays to 'photograph' atoms worked especially well for crystals, where the atoms are spaced regularly in a lattice. Knowledge of the wavelength enables an estimate of the sizes of atoms to be made, which turns out to be about 10^{-10} m, a value confirmed by other techniques. Knowing the size of the spacing, the numbers of atoms in a given quantity of material can be calculated. A kilogram of coal (mainly carbon) contains about 5×10^{25} atoms, hence each atom of carbon weighs roughly 2×10^{-26} kg. The lightest atom is that of hydrogen, which weighs about 1.7×10^{-27} kg. However, historically, the first truly accurate information about sizes and weights of atoms came from studying their electrical properties (see below).

A substance which consists of just one type of atom, perhaps in molecular groupings, is called an *element*. There are many different types of atoms, and hence many different elements. After the lightest atom, hydrogen, the next lightest is helium, then lithium, and so on up through carbon, oxygen, magnesium to iron, and then to the very heavy atoms such as lead and uranium. On Earth there are about 90 naturally-occurring elements, and a few heavier ones have been made artificially. Although they all differ in weight, many atoms have similar chemical properties; for example, sodium and potassium react with many substances in the same way. In Section 3.1, when something about the internal structure of atoms has been explained, more will be said about the fact that many different types of atoms exist.

2.2 *The internal structure of atoms*

There is much evidence which suggests that atoms are *electrical* in nature. One of the most persuasive early experiments on the relationship between atoms and electricity was conducted in the early nineteenth century by Faraday, who was interested in the phenomenon of electrolysis – the decomposition of a chemical compound contained in a liquid solution into its constituent elements, which can be brought about by the passage of an electric current. Using electrolysis, it is easy to split molten common salt (sodium chloride) into sodium and chlorine, for example (see Fig. 2.2). The significant fact which Faraday discovered is that a fixed quantity of electricity is always required to electrolyse a fixed weight of elements; which is to say a certain number

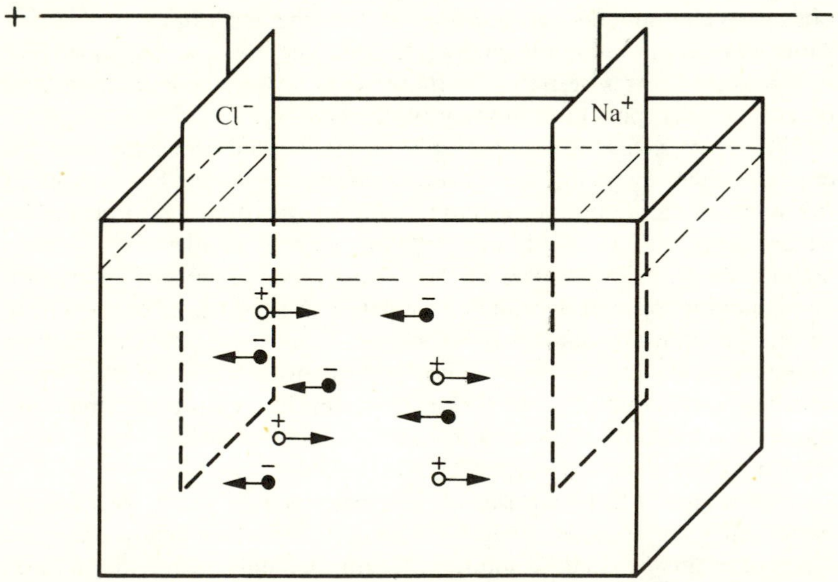

Fig. 2.2. Atoms are electrical. When two electrodes (electric plates or terminals) are inserted into molten common salt (NaCl) the electric field separates the sodium (Na^+) which drifts to the negative electrode from the chlorine (Cl^-) which accumulates at the positive terminal. This suggests that the sodium and chlorine atoms are electrical and stick together by electric forces to form salt.

of atoms. Twice the electricity gives you twice as many atoms. This suggests that electricity comes in lumps – definite units of charge – and that each charge attaches itself to an individual atom. When an electric field is applied to the solution, these charged atoms – or *ions* as they are called – are attracted by the electric field towards the positive or negative electrode depending on the sign of the charge, and there deposit themselves. In this way the separation of a chemical compound into its constituent elements can be achieved. When molten salt is electrolysed, the chlorine ions appear at the positive electrode, suggesting they carry a negative charge, while the positively charged sodium turns up at the negative electrode.

Furthermore, the *weights* of different materials deposited by the same given quantity of electricity were found by Faraday to bear the same ratios as the weights which combine together chemically, which is precisely what one would expect if each electric charge deposits one atom. These early facts of electrolysis already provided highly convincing evidence for the existence of atoms, with each chemical element containing identical atoms of a fixed weight which can become charged with electricity, the charge having a negligible weight.

In spite of this, it is known that atoms in their normal form are not electrically charged, so the charges must be able to detach or attach themselves to the atoms according to circumstances. The precise relationship between electric charges and atoms became clear with the work of the physicist Joseph John Thomson (British, 1856–1940).

In Chapter 1 it was explained how electricity behaves rather like an invisible fluid that can flow along a wire as a current, or be stored in an insulated metal object. Nothing was said about the vehicle on which the electricity resides and travels. Although the properties of electricity and magnetism were well understood by the 1890s, very little concrete information had been obtained on what it actually is, how it moves about and how it can appear and disappear.

The study of electric currents was a major preoccupation of experimental physicists in the nineteenth century. It had long been established that electricity flows through certain solid materials like metals, and also through some liquids, including water. Thomson was studying the flow of electric current through *gases* (this is the principle of the fluorescent tube light), a discovery attributed to Julius Plücker (German, 1801–68) and much studied by Sir William Crookes (British, 1832–1919). The flow is particularly efficient when the gas is very tenuous (at a low density and pressure): the electricity seems to have no difficulty in leaping across the empty spaces between the gas molecules. Indeed, by pumping down the pressure sufficiently, Thomson was able to study a concentrated beam of almost pure electric current. By allowing the beam to strike a fluorescent screen, in a sort of precursor of the television tube, Thomson was able to keep track of where the beam was.

Many people had experimented with these cathode rays (so named because they emerge from the negative electrode, or cathode) but their precise nature remained something of a controversy until Thomson's work. The 'rays' can easily be bent by a magnetic field, suggesting a negative electric current, but there was disagreement about whether they were some sort of waves, like light, or material particles with a definite mass. What Thomson did was to establish that the rays were indeed tiny charged particles (or 'corpuscles' as he called them) by measuring reasonably accurately their mass to charge ratio, m/e, an endeavour for which he received the Nobel prize in 1906.

He did this by applying known electric and magnetic fields to a beam of the rays, and measuring how strongly the beam could be bent by these forces. The inertia (mass m) of the particles tries to make them continue in straight line motion, but the charge (e) feels the electric and magnetic fields which try to deflect their motion. By studying the competition between these influences, the ratio m/e is determined (see Fig. 2.3).

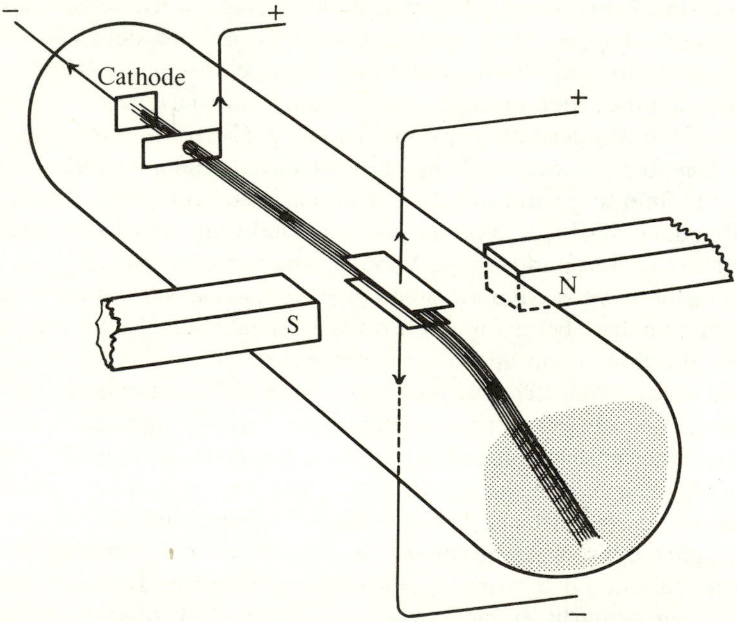

Fig. 2.3. The first subatomic particle. Schematic diagram of J. J. Thomson's apparatus to measure m/e for the electron. A beam of 'cathode rays' in a low-pressure gas discharge tube passes between electrically charged plates and the poles of a magnet. By varying the field strengths and measuring the deflections of the beam as it strikes the fluorescent screen the mass to charge ratio can be determined.

Now the earlier work of electrolysis had enabled the ratio of mass to charge for *atoms* (strictly speaking ions) to be measured, simply by noting how much electricity is needed to decompose a given mass of a chemical element. The smallest value (lightest atom) was obtained for hydrogen. Assuming that the charges were the same, Thomson's corpuscles turned out to be more than 1000 times lighter than the hydrogen atom. Thomson had discovered the first subatomic particle. The name which was adopted for these 'atoms' of negative electricity is *electron*, coined some years previously by an Irish physicist George Johnstone Stoney (1826–1911) who had proposed the existence of microscopic electric particles.

Neither the experiments in electrolysis, nor Thomson's work had provided a measurement of the actual mass of an atom, or the electron, itself; only the ratio m/e. Some crude estimates of atomic masses had been made by studying surface tension (see p. 93) and other phenomena, but knowledge of m/e could be used to obtain an accurate value only if the charge e could be independently measured.

This was achieved in a series of experiments culminating in the work of the American physicist Robert Andrews Millikan (1868–1953) in Chicago, first presented in 1909. The method adopted was to produce a fine spray of water (later oil) droplets between two large electrically charged plates. A nearby source of X-rays or radium was used to ionize the air and enable the droplets to pick up individual electric charges. The electric field between the plates, acting on these charges, had the effect of buoying up the droplets against their own weight. The weight of a droplet can be measured by switching off the field, letting the droplet fall, and studying its progress against air resistance. Millikan found that the electric charge on the droplets was always a definite whole number multiple of a fixed unit, which he took to be the charge of a single electron. The answer came out to be very small. One way of visualizing it is to note that a current of about one ampere (a typical current for a domestic appliance) represents the passage of no less than 10^{19} electrons per second past a fixed point in a wire.

Since the work of Millikan, who was awarded the 1923 Nobel physics prize, all experiments have confirmed that the basic unit of electricity in nature is numerically equal to the charge on the electron. Knowing e, the electron mass can be calculated from m/e and works out at 9.1×10^{-31} kg, the mass of the hydrogen ion being about 1836 times larger.

The fact which had to be explained was, if ordinary atoms are electrically neutral, where do the electrons come from? Clearly they must be hidden inside atoms along with positively charged particles of some sort, so that overall electrical neutrality can be achieved. An atom may then become ionized (charged) by losing or gaining an extra electron. In fact, it became possible around the turn of the century to produce beams of these ions in variants of the cathode ray tubes, and to measure their *m/e* ratios directly, confirming that they were the same as the ions deposited in electrolysis. Knowing *e*, these techniques of electric and magnetic deflection led to machines capable of accurately measuring atomic masses.

Besides the cathode ray tube, there are many other ways of extracting electrons from ordinary matter – bombardment by radiation such as X-rays, shining light on to metals (the photoelectric effect), heating metals, even ordinary friction. In contrast, these techniques do not produce *positively* charged subatomic particles, except indirectly by leaving behind a positive ion when an electron has been removed. The deflection experiments show that positive charges tend to be found stuck permanently to the major bulk of the atom, whereas the light, mobile electrons can be detached fairly easily.

The fact that positive charges are found attached to heavy, ponderous particles is confirmed by an entirely different set of phenomena, called radioactivity (to be discussed in detail in Chapter 3). Certain very heavy atoms such as uranium are unstable, and eject heavy, positively charged particles spontaneously. When these radioactive emissions were first discovered by Antoine Henri Becquerel (French, 1852–1908), winner of the 1903 Nobel physics prize, they were called alpha rays, but are really *particles*. Measurements with alpha particles soon revealed that they carry a positive charge whose magnitude is exactly twice the negative charge carried by the electron, though they are more than 7000 times as heavy as the electron.

With the discovery of alpha particles, physicists soon came to understand that atoms are not solid, indivisible bodies after all, but are themselves composed of smaller electric particles bound together by electric and magnetic forces. Intense interest then surrounded the question of the *internal* structure of the atom. What did it look like? How were the electrons and heavy positively charged particles arranged? What prevented the electrons simply sticking to the positive charges and neutralizing them?

The internal structure of atoms was first proposed by the New Zealand physicist Ernest Rutherford (1871–1937) following experiments on the scattering of alpha particles by metal foils. The idea behind his experiment, which was the precursor of what is still the most important technique for obtaining information about subatomic systems, is to use microscopic projectiles as *probes*. In his case, high-speed alpha particles were allowed to impinge on a variety of substances, and their fate monitored as they emerged from the other side. To keep track of the particles, Rutherford placed a fluorescent screen behind the target and looked for the tiny flashes which indicate that an alpha particle has struck the screen.

The situation can be compared to firing bullets at tissue paper – the alpha particles possess so much energy that they can easily plough through the foil without being stopped. It might then be expected that a beam of them would simply penetrate the foil unhindered and hit the screen on the far side. What Rutherford actually found was that the alpha particles spewed about at all angles, some of them even bouncing backwards off the foil. He reasoned that the tiny projectiles were being *scattered* by electric charges inside the foil. The observed pattern of scattering can be explained as *ricochets* which occur when the alpha particles collide with something rigid.

Rutherford postulated that as the particles travel through the metal they encounter millions of atoms. The alpha particles are so small compared to the size of the atom (about one ten thousandth or less) that they can move right through the *interior* of the atom and probe its internal electric fields. If the atoms really do contain electrons and positive charges then the alpha particles will come under the action of their electric fields and be deflected.

To explain the large amount of scattering observed, Rutherford proposed a model of the atom in which most of the mass is concentrated into a compact body called the *nucleus*, with the lighter, mobile electrons surrounding it in a very distended and tenuous cloud. All the atom's positive charge is located in the nucleus, and it is the electric field of the nucleus (not the nucleus itself) which scatters the particles. The electric force is given by Equation (1.4). Because most nuclei are considerably heavier than the alpha particles, they do not suffer much recoil under the impact of the bombardment, so they behave almost like rigid, fixed lumps distributed throughout the metal foil. (The insubstantial electrons are easily swept aside under the onslaught of the alpha battering.)

As the nucleus is so compact most of the alpha particles will pass by at a distance where the nuclear electric field is weak and cannot impart much of a deflection. Occasionally, however, a head-on collision will occur and the alpha particle will approach the nucleus closely. The repulsion between the nuclear charge and that on the alpha particle will build up until the force is strong enough to halt the charging alpha particle and reverse it in its tracks, propelling it back along its incoming route and seemingly causing it to 'bounce back' off the foil.

Knowing the energy of the alpha particles Rutherford could calculate how close they could penetrate to the nucleus during head-on collisions. He found it to be within 10^{-14} m. Even at this proximity they behaved as though they were moving through empty space in a pure electric field, showing no evidence of actually striking the nuclear surface. Evidently the size of the nucleus is very small indeed: no larger than 10^{-14} m, or one ten-thousandth of the size of the atom as a whole. This implies that nuclear matter is incredibly dense: at least 10^{18} kg m^{-3}, or one million billion times the density of water.

In the Rutherford atom the electron cloud must be prevented from collapsing on to the nucleus under the attraction of the intense electric field that surrounds it. The complete explanation of this stability had to await the inception of quantum mechanics in the 1920s, but a partial explanation immediately suggests itself. The atom is structured rather like the solar system, where the planets avoid falling into the sun under the attraction of its gravitational field by orbiting around it at great speed, thereby balancing the force of the sun's gravity with their centrifugal force. Similarly, the electrons fly around the nucleus in tightly curved orbits at a fraction of the speed of light, and centrifugal effects prevent their migration towards the nucleus.

The general picture of the atom which Rutherford envisaged is now universally accepted, and he received a Nobel prize in 1908. In the succeeding years the internal arrangement and structure of atoms have been intensively studied using a variety of advanced technological probes, so we now have a very clear and detailed picture of the atomic architecture. In particular, we can easily understand why there are so many different *types* of atoms (i.e. many chemical elements). As the charge on the electron is always the same basic unit, and as all atoms in their normal condition are electrically neutral, it follows from the planetary model that the nucleus must contain an equal quantity of positive charge. Thus, the nuclear charge always comes in *fixed multiples* of a basic unit, with the same magnitude as the electronic charge, but the opposite sign. The chemical and physical differences between atoms are due to the different number of electrons that they possess, which in turn is due to the different values of their nuclear charges.

The simplest atom is that of the element hydrogen, and consists of a nucleus with just one unit of charge, and has one attendant electron to balance it. Hydrogen is also the commonest of all atoms: about 70% of the universe is made of it. The next atom is helium which has a nuclear charge of two units, and hence two electrons. The helium nucleus is in fact none other than the alpha particle. Helium is also the next commonest of atoms: nearly all the remaining 30% of the universe is made of helium.

It is straightforward to continue down the list of atoms consecutively, according to their nuclear charges: lithium, beryllium, etc., on through carbon, nitrogen, oxygen, to iron, gold, lead and finally uranium. These constitute 92 types of atom in all. Thus, uranium has a nuclear charge of 92 units, and possesses 92 electrons. Furthermore, the highly charged nuclei are also the heavy ones. As mentioned, there are some atoms heavier than uranium which have been made artificially. These contain more than 92 units of nuclear charge. All the atoms between hydrogen and uranium are known, so there is no room for the science fiction idea of an unknown element.

The electrons in the swarm around the nucleus do not move about independently but interact electromagnetically with each other in a highly complex way. In an atom of uranium there is an immense network of changing forces as the dozens of electrons whirl around encountering and repelling each other with electric forces. Their motions also set up complicated magnetic fields which give rise to still more forces. At first sight it might appear that such an overcrowded and competitive environment must disintegrate into anarchy. Remarkably, however, even the heavy atoms maintain a highly orderly and disciplined arrangement. In fact, in the nineteenth century the Russian chemist Dmitri Ivanovich Mendeleev (1834–1907) had already shown that the many chemical elements could be organized into a systematic pattern known as a periodic table, in which elements with closely similar chemical and physical properties (e.g. sodium and potassium) are grouped together. The structure of the periodic table is now broadly understood in terms of the ordered arrangements of the orbital electrons of the different kinds of atoms.

To appreciate this, note that chemical processes occur when whole atoms interact with each other, such as when they combine to form molecules. The nature of *interatomic* forces can only be fully understood with the use of quantum mechanics (see Section 2.4) but are essentially electrical in origin. If two atoms approach, their complicated electric fields come into contact with the other's electrons and the combined system becomes disturbed. Only the outermost electrons are seriously affected by this, because they are rather weakly bound due to the fact that they are shielded from the attraction of their nucleus by the repulsive fields of the swarms of more tightly bound electrons. These outer electrons rearrange themselves under the action of the disturbing fields from the other atom, and may even desert the parent atom altogether and take up residence around *both* atoms at once, orbiting round the combined system. The effect of this rearrangement is to redistribute the electric charges associated with the atoms, and if this results in a somewhat lower overall electric and magnetic energy, then the combination is energetically favoured and will persist. The two atoms will then be bound rather loosely together to form a molecule. Many atoms can unite in a single grouping this way, perhaps forming regular shapes such as tetrahedra, rings and helices, or long tangled chains.

Clearly the details of the interatomic chemical forces will depend in a very complicated way on the arrangement of the outermost electrons round the atoms. In particular the *number* of electrons which an atom possesses will be an important factor in determining its chemistry. But this number is fixed by the electric charge on the nucleus, for the simple reason that if a nucleus has a charge of, say, 20 units, it will pick up 20 electrons from the surrounding world; then it will be electrically neutral. Thus it is the nuclear charge that ultimately determines the chemistry of the atom. All the elements long ago identified by chemists by their different chemical properties are seen to differ from each other essentially because of the different values of their nuclear charges. It is the nuclear charge that makes iron chemically different from sulphur, or tin, or helium.

The planetary atom also explains where electricity, so ubiquitous and easily produced in the world around us, comes from. As explained above, the outermost electrons are not particularly strongly bound to the atom because of the shielding of the nuclear electric field by the inner electrons. This binding can become still weaker if the atom comes under the action of other external forces, such as the presence of another atom nearby. For example, in a crystal lattice atoms are locked together by fairly strong interatomic forces into a regular array. Under these circumstances electrons may easily leave the proximity of any particular atom and wander about in the crystal. Some metals have a crystalline structure which contains an abundance of these liberated peripheral electrons, and when an electric field is applied to the metal the electrons flow along the field lines and form a current. In this way electric currents are explained as an orderly migration of discarded electrons.

As far as we can tell there are the *same* number of electrons inside ordinary matter as there are positive units of nuclear charge. Just how or why the universe has contrived to balance the electrical books in this way is still a mystery.

2.3 *Probing the atom with light*

With the understanding that atoms contain moving electric charges, it comes as no surprise that they can emit electromagnetic radiation. Indeed, the fact that hot objects glow with heat and light radiation is another indication that ordinary matter contains electric particles somewhere, which are disturbed by the heat energy. In Section 1.4 it was explained how electromagnetic radiation is generated by rapid accelerations of charged particles, so the possibility arises of studying the motions of the charges in atoms by analysing the quality of the light they emit. More than anything else, this technique has provided detailed information about atomic structure.

One reason for this is that measurements carried out on *waves* can be exceedingly accurate due to the phenomenon of *interference*. This familiar effect occurs when two or more waves come together so that their motions combine, a situation known as *superposition*. If two wave trains with the same wavelength intersect obliquely, where the crests of the waves coincide, the wave disturbance is reinforced, but if the crests of one wave coincide with the troughs of the other, some cancellation of the disturbance occurs. By experimenting with interference patterns physicists can measure the wavelength of any light extremely accurately. Furthermore, knowing the speed of light, the frequency of the light wave can also be deduced from Equation (1.5).

The relevance of the frequency and wavelength of light radiation to the internal motion of atoms is easily recognized once it is remembered that a particle which travels along a curved path is *accelerating* (see Section 1.1). If for simplicity we imagine an electron in an atom to be orbiting around the nucleus along a circular path, then its acceleration is equal to $\omega^2 r$, where ω is the angular frequency of its rotation and r the radius of the orbit. Because it is accelerating, we would expect this electron to emit electromagnetic waves. Now it is easily shown mathematically, and is obvious intuitively, that if an electron rotates around the atom $2\pi\omega$ times a second, then the frequency of the electromagnetic wave emitted will also be $2\pi\omega$. This tells us that the electrons in the atom must be moving very fast, because visible light has a frequency of around 10^{15} Hz, so the electron apparently makes 10^{15} atomic revolutions in one second. For a typical atom, the radius $r = 10^{-10}$ m, so we conclude that the speed of the electron is about 10^6 m s^{-1} – considerably faster than the speed at which the Earth orbits the sun, but still much slower than light (10^8 m s^{-1}) which indicates that, to a first approximation at least, we can neglect relativistic effects in considering the emission of light by atoms.

Although we now know that the behaviour of microscopic particles is considerably more subtle than has been described so far, to the physicists of the early 1900s the image of a miniscule 'solar system' with tiny electric particles rotating about a heavy nucleus, emitting electromagnetic radiation as they went, seemed very real and it was natural that they should construct a mathematical model of this system by applying Newton's laws of motion and Maxwell's theory of electromagnetism. The ingredients of such a model are extremely simple: electric forces attract the electrons to the nucleus, and the electrons avoid being drawn into the nucleus by circling around very fast and balancing the electric attraction with their outward centrifugal force. The nearer the electron orbits to the nucleus the greater the force (recall the inverse square law of electric force from Equation (1.4)) so the faster it must revolve to counterbalance this attraction (just as Mercury, being close to the sun, orbits it in 88 days while the more distant Earth takes 365 days). For a circular orbit, the equation describing this balance is simply electric force = centrifugal force or

$$eQ/r^2 = m\omega^2 r \tag{2.2}$$

where e is the electron charge, Q the nuclear charge and m the electron mass. This equation can also be read as Newton's second law: force = mass × acceleration. Rearranging the formula gives

$$\omega^2 = eQ \div mr^3 \tag{2.3}$$

which shows that as the orbital radius r decreases, and the electron approaches the nucleus, the rotational frequency goes up.

An investigation of the energy of an atom is also instructive. The electric attraction of the nucleus tries to draw the electron towards the centre at $r = 0$ so to remove an electron from the influence of the nucleus requires an *input* of energy to pull it free. The electric energy is therefore negative, and equal to $-eQ/r$. On the other hand, the circular motion of the electron tries to pull it away from the nucleus, so this energy of motion is positive, and equal to $\frac{1}{2}mv^2$ where v is the velocity of motion. The relation between circular and ordinary velocity is $v = \omega r$, so the kinetic energy is also equal to $\frac{1}{2}m\omega^2 r^2$. There is thus competition between electric and kinetic energy, with the net energy given by $\frac{1}{2}m\omega^2 r^2 - eQ/\mathrm{r}$. We can use Equation (2.2) to eliminate $m\omega^2$, and write the net energy as

$$\text{energy} = -eQ/2r. \tag{2.4}$$

This result is negative because the electron is *trapped* by the nucleus – the kinetic energy is insufficient to break the electric bond to the nucleus. To free it, a quantity of energy equal to (2.4) must be supplied to the atom. If an electron orbits rather close to the nucleus (small r) then Equation (2.4) tells us that the energy of binding is very strong. If some energy were pumped into the system, causing the electron to move faster, it would take up a new orbit farther from the nucleus (large r). Thus, the tightly-bound orbits have a large negative energy and small radius, while the loosely bound orbits have small binding energy and large radius. Furthermore, the angular frequency of revolution ω is highest for the tightly bound electrons, and steadily diminishes to zero when the electron moves at very great distance from the nucleus.

Let us consider the interaction of atoms with light in more detail. For clarity suppose we are dealing with the simplest of all the atoms; that is, hydrogen. Each atom of hydrogen consists of just two particles, a heavy nucleus, and a single light electron which orbits around it. Suppose we have a quantity of hot hydrogen gas, consisting of many billions of these atoms, all rushing around chaotically and colliding with each other through their electromagnetic forces. Each atom receives and delivers energy via collisions with other atoms, and also by absorbing and emitting electromagnetic radiation in the form of heat and light. Thus, due to these random processes, the available energy will be distributed among the atoms in a continuous way, some having rather more than others. Naturally some of the energy is communicated to the electrons also, and those that have a rather large share become less tightly bound to the nucleus (with large orbits) while others occasionally lose some energy and spiral in close to the nucleus. In an average selection of atoms we would expect to find all sizes of orbits (values of r) and hence all values of the frequency ω (see Equation (2.3)). So one would expect that if we were to examine the light radiation from all these atoms, we would find all types of frequencies (colours) mixed up together.

The instrument used for analysing the frequency mixture of light is called a spectroscope. Its principle is a simple one. Everybody is familiar with the phenomenon that when a straight stick is held obliquely through the surface of a pond, it appears bent. This is because light travels more slowly in the dense water of the pond than in the air. A closer investigation reveals that short wavelength light is slowed down more than long wavelength light. This provides a means of separating out the various wavelengths by passing the light through a prism. The different wavelengths are then bent by different amounts and can be directed to different regions of a photographic plate. Recalling that the frequency (or wavelength) of light corresponds to the colour quality, the effect of passage through a dense object such as a prism is to separate the light into its constituent colours; this spectrum of light is what is observed in a rainbow, when the spherical raindrops play the role of the spectroscopist's prism.

Fig. 2.4 shows the spectrum of light from hydrogen gas, and it comes as something of a surprise. According to the atomic model described above, we would expect to see a continuous 'rainbow' of colours corresponding to all the atoms emitting light with all their different and

Fig. 2.4. Hydrogen spectrum, showing an ordered pattern of discrete lines.

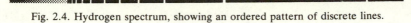

various orbital radii and frequencies. Instead, only certain sharply defined frequencies appear. Clearly there is something fundamentally wrong with the simple planetary model of the atom. The discrete lines, which are characteristic of all atomic spectra, not just that of hydrogen, indicate that certain frequencies, and hence certain definite atomic orbits, are favoured over others, a mystery that required nothing less than a total revolution in our understanding of the physical world to explain.

Besides the enigmatic nature of atomic spectra, theoretical physicists were greatly puzzled by a *mathematical* inconsistency in the planetary model of the atom. The problem concerns the fact that light carries energy. When an electron whirls around in an atom, giving off light, it loses energy. But, as we have seen, the lower (i.e. more negative) the energy of the electron, the closer it is to the nucleus, and the faster it rotates. This in turn means that it radiates energy still more vigorously, causing its orbit to shrink further. It seems an inescapable conclusion that the orbits are actually unstable and the electrons would spiral inexorably inwards, causing the atom to collapse. A careful study of the statistical balance between the energy of electromagnetic radiation and the energy of electrons leads to the result that in a cavity filled with radiation at the same temperature as the walls of the enclosure, the energy would steadily flow out of the walls and into the radiation field with the unhappy consequence that equilibrium could only be reached when the energy of the radiation is infinite!

With the realization that atoms were unstable it became clear that there was a fundamental misconception about the nature of microscopic matter. The first step in the resolution of these problems was taken by the German physicist Max Karl Ernst Ludwig Planck (1858–1947, Nobel prize 1918) in 1900. He was able to show mathematically that the imbalance of energy between radiation and matter at a common temperature could be alleviated if the energy which is transferred from one to the other always arrives or leaves in definite packets or lumps, which he called *quanta*. The energy of a quantum of radiation, such as light, is related to its *frequency*, ν, through a universal formula

$$E = h\nu \tag{2.5}$$

where h is a constant known, appropriately enough, as Planck's constant. Its value is 6.6×10^{-34}J s: this means that a quantum of energy is very small indeed. A quantum of visible light ($\nu = 10^{15}\ \mathrm{s}^{-1}$) is only about 10^{-18} J, so an average domestic light bulb will emit 10^{20} quanta each second. Equation (2.5) may also be written in terms of the *wavelength* of light using Equation (1.5)

$$E = hc/\lambda\,. \tag{2.6}$$

Using the quantum radiation hypothesis, not only was Planck able to remove the serious inconsistency about the thermal balance between radiation and matter, but he was actually able to compute the precise spectrum of the radiation inside a cavity which has reached thermal equilibrium. This characteristic spectrum is beautifully confirmed by direct experiment (see Fig. 2.5).

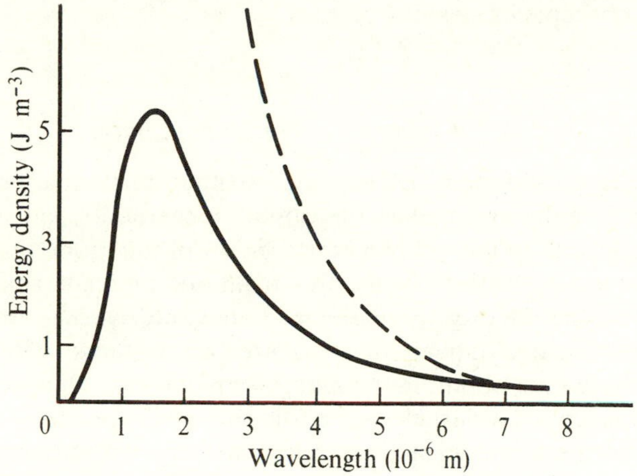

Fig. 2.5. Planck's radiation spectrum, for a body in thermal equilibrium at 2000 K. The broken line shows the results of the old theory, which predicts a catastrophic output of energy at the short wavelengths (high frequencies).

Quanta of light are often called *photons* and have some of the characteristics of particles. Good evidence for the quantum nature of light comes from the photoelectric effect mentioned on p. 44. Recall that this is a phenomenon which occurs when light falls on the surface of certain metals and knocks electrons out of it. Experimentally it is found that a more intense beam of light will liberate more electrons at the same energy rather than delivering more energy to the same number of electrons. Turning up the frequency, however, does not change the number of electrons which are displaced, but does increase their energy. This was explained, using the quantum theory of light, by Einstein in 1905, for which, along with his theoretical work, he was awarded the 1921 Nobel prize in 1922. Envisaging the electrons in the metal as under bombardment by discrete photons, the displacement of an individual electron can be explained by supposing that it absorbs the energy of a *single* photon, thereby breaking free of its atomic binding. Increasing the intensity of the light simply increases the flux of bombarding photons, resulting in the liberation of greater numbers of electrons. But increasing the frequency does not change the *flux* of photons, it just gives them more energy (see Equation (2.5)) to knock out more vigorously the same number of electrons.

The quantum hypothesis has turned out to have the most profound implications for our understanding of the world in general, and the nature of matter in particular. Quantum theory now encompasses not just the behaviour of radiation but the whole of microscopic physics, and has produced a picture of nature which is far removed from the intuitive, everyday concepts of classical physics.

2.4 *The inherent uncertainty in matter*

So far we have used the word 'particle' without much qualification, but what does it really mean when the object concerned is too small to be seen or touched? In the old planetary model of the atom, the electrons were regarded as entities which differ from our everyday experience of particles (such as the proverbial billiard balls) only by *scale*. But how can we be sure that qualitatively new features do not arise at microscopic dimensions that are absent in the macroscopic world? How do we know that the concept of a particle is meaningful at all for such a tiny size? These problems are relevant quite independently of whether an electron is considered to have a certain structure or extension, or whether it is considered as a point-like object, with all its material concentrated at a single place.

Building on the foundation of Planck's idea of the quantum of radiation, Niels Henrik David Bohr (Danish, 1885–1962) suggested in 1913 that the quantum hypothesis should be applied to matter, as well as light, in an attempt to model the structure of the atom. He proposed that just as the energy of a photon is restricted to certain fixed multiples by Equation (2.5), so should the energy of electrons in an atom be similarly restricted. Applying this idea to a planetary model of the atom, Bohr was able to reproduce the essential feature of atomic spectra – the series of discrete lines – by assuming that photons are emitted when electrons make sudden jumps between these atomic energy levels. The frequency of these photons will then be restricted, via Planck's relation (2.5), to the fixed energy gap between two levels. For a simple atom, such as hydrogen, Bohr's model gives in outline the correct formula for the frequencies to a good accuracy. He was awarded a Nobel prize for this in 1922.

An understanding of *why* the electrons in an atom should obey quantum rules of the kind suggested by Bohr had to await the revolutionary new theory of matter developed in the 1920s by Erwin Schrödinger (Austrian, 1887–1961) and Werner Karl Heisenberg (German, 1901–76) called quantum mechanics. According to this theory, it is not merely the *laws* governing the motion of microscopic particles like electrons that are different from the laws of motion for billiard balls, but the entire conceptual foundation. A full understanding of the new theory is very difficult to achieve, and may only be given proper expression through the language of advanced mathematics. Many of the consequences of the theory are bizarre and have an Alice-in-Wonderland aspect. Although total understanding is unnecessary for the contents of this book, the forces and fields which control the microscopic structure of matter operate according to the principles of quantum mechanics, so something of its general flavour must be grasped before these developments can be properly understood.

Let us begin by deciding what characteristics are possessed by a particle of matter in the macroscopic world (think of a billiard ball). First, it is regarded as having a precise location; that is, it is always centred on a certain place. Also, it will have a definite motion; in technical language we can assign to it a certain momentum. In addition to these external qualities it may also possess internal properties: it could, for example, be spinning. When people think of electrons whirling around in atoms, they tend to imagine these as objects at a definite place moving in a definite way. Useful though this mental image may be, the electron is so small, how can we really know?

How can one find where an electron is? Recall the problem of seeing things with waves, mentioned on p. 39. To resolve the detail of a small object of a certain size it is necessary to illuminate it with waves, the wavelength of which is no larger than the size of the object. X-rays are fine for looking at atoms, but the electrons are supposed to be much, much smaller (perhaps even no size at all). This requires much shorter wavelengths, say gamma rays. Whatever waves one uses we can only locate the electron to within a region greater than at least one wavelength, so the greater the precision of location required, the higher the frequency (shorter the wavelength) needed.

Consider what happens when these impinge on the target electron. The electron will not remain inert, but will suffer a recoil as the electromagnetic wave (which carries energy and momentum) disturbs it. The reason for the recoil is easily explained. The electric field of the wave accelerates the charge in its direction, so the charge becomes a tiny electric current. The magnetic field of the wave then acts on this current with a magnetic force, just as a magnet exerts a force on a current-carrying wire (see p. 18). An examination of the direction of the fields in the wave shows that the resultant force delivered to the electron is in the direction of the motion of the wave.

It follows from this that in order to locate an electron, it is inevitable that the electron itself gets knocked out the way – its motion is disturbed. Thus, if we wish to know the position, we introduce an uncertainty in the motion (momentum) of the particle. The more energetic the incoming wave, the greater the disturbance. Now it might be thought that the recoil could be reduced as much as one pleases by reducing the wave energy, but it is here that the quantum property of the photon, as proposed by Planck and exemplified by Equation (2.5), plays a significant part. According to that fundamental equation, the energy of light of a fixed frequency ν *cannot* be reduced below $h\nu$ (except, of course, by having no light at all). Because light comes in lumps (quanta) we must use at least one lump to see the electron. But the energy of the quantum is proportional to its frequency, so to get the energy down (reduce E) it is necessary to reduce the frequency also; i.e. to increase the wavelength. It follows that the requirements of high-precision location and low recoil are mutually incompatible. We can either have a low-energy wave which does not disturb the motion of the electron very much, but the wavelength of which is large, thus leaving a high uncertainty in the position measurement, or we could use a short-wavelength, high-energy wave and fix the position accurately at the expense of the motion. Either way there will be a residual uncertainty somewhere.

The relationship between the uncertainties in position and momentum obey a simple formula

$$\Delta x \Delta p \gtrsim h \tag{2.7}$$

where Δx, the uncertainty in position, times Δp, the uncertainty in momentum, is greater than about the value of Planck's constant h. (A more precise analysis fixes $\Delta x \Delta p$ to be greater than $h/4\pi$.) It is obvious from (2.7) that as Δx is increased, Δp is decreased and vice versa. This relationship is now known to be a fundamental property inherent in all matter. It represents more than just a limitation to our technological capabilities in measurement; it is an elementary feature of nature.

Thus, it is meaningless to attribute, even in principle, a definite position simultaneously with a definite momentum. These concepts are complementary – we can either have one, *or* the other, but not both. If an electron is trapped in a tiny box, so we *know* it is confined to a certain small region of space, then we can know almost nothing of its motion inside the box. Alternatively if we shoot an electron through some apparatus in such a way that it can only get through if it has a definite velocity, then we can know nothing of its position in the apparatus. The relationship (2.7) therefore plays the role of a basic principle of physics, and is known as Heisenberg's uncertainty principle.

The Heisenberg principle applies to all particles of matter, not merely electrons. Indeed, it even applies to billiard balls. However, the smallness of h means that the effects of this inherent uncertainty are only important in microscopic systems such as atoms; they are totally negligible in everyday objects. To give an example, if we were to measure the position of a 1-kg ball to an accuracy of one atomic radius (10^{-10} m) we would introduce and uncertainty in its velocity of a miniscule 10^{-23} m s^{-1}.

The Planck relation (2.5) leads directly to another uncertainty relation of the type shown in Equation (2.7). Suppose we wish to measure the energy of a photon by measuring its frequency ν and using (2.5). To measure ν it is necessary to allow the electromagnetic wave to oscillate at least through one cycle, which takes a time of ν^{-1}. If the time is uncertain by $\Delta t \simeq \nu^{-1}$, then the energy is uncertain by at least an amount ΔE which is some large fraction of E. But $E = h\nu$, so $\Delta E \simeq h\nu \simeq h(\Delta t)^{-1}$ which gives

$$\Delta E \Delta t \gtrsim h. \tag{2.8}$$

This uncertainty relation tells us that we cannot simultaneously measure the energy and the time. It turns out to apply not only to photons, but to all subatomic systems, and leads to phenomena of immense importance. For example, (2.8) tells us that for a short duration the law of energy conservation, one of the most fundamental principles of classical physics (see Section 1.1), can be violated. It is possible for a system, such as a subatomic particle, to 'borrow' some energy ΔE, so long as it pays it back before an interval $\Delta t \simeq h/\Delta E$ has elapsed. In later sections we shall see that this possibility of 'energy credit' profoundly changes the behaviour of subatomic matter.

2.5 *Waves and particles*

In the previous section it was explained that there is an inherent uncertainty in the condition of microscopic particles. This conclusion does not rest alone on the analysis of the interaction of light with electrons: we could, for example, use other *particles* to try and measure the position and motion of electrons by scattering them from the target electron at very low energies so as not to induce too much recoil.

However, this strategy is also doomed to failure because of another unexpected property of microscopic systems, equally as remarkable and significant as Planck's quantization of radiation. Planck's idea invests wave fields with a sort of particle-like property, requiring photons of a definite energy and momentum to interact with matter. In 1924 the French physicist Prince Louis Victor de Broglie, who was awarded the Nobel prize for physics in 1929, suggested that just as waves acquire particle properties on a microscopic scale, so should particles of matter display wavelike properties. Indeed, he proposed that a particle of momentum p should be associated with a wavelength λ related through the equation

$$p = h/\lambda \tag{2.9}$$

where h is the *same constant* as appears in Planck's quantum relation (2.5). This extraordinary and brilliant idea of a wavelike aspect to matter is a totally new conception of the nature of subatomic particles like electrons, which were previously pictured as simply scaled-down 'billiard balls'.

De Broglie's theory had some support from a series of experiments carried out during the early and mid 1920s in the United States under the direction of Clinton Joseph Davisson (1881–1958). These experiments involved the study of the scattering of a beam of electrons reflected from the surface of a nickel crystal. The atoms of a crystal are arranged in a very ordered pattern, which is reflected in the pattern of the scattered electrons. What Davisson and his colleagues discovered, especially in an improved version of the experiment conducted in 1927 with Lester Halbert Germer (American, 1896–1971), was that the pattern of the scattered electron beam was exactly the same as one would obtain if the beam were not composed of particles at all, but of waves. Specifically, the electron beam displayed the phenomenon of *interference* (see p. 49) such as one obtains when wave patterns merge together. The *length* of the waves, λ, was found to be inversely proportional to the *momentum*, p, of particles in exact accordance with de Broglie's simple relation (2.9). These experiments, together with similar ones involving the scattering of high-energy electrons from thin metal foils, carried out by George Paget Thomson (1892–1975), the son of J. J. Thomson, provided direct evidence of the wave nature of matter. Davisson and Thomson shared the 1937 Nobel physics prize.

In spite of the surprising idea of electrons behaving like waves, the phenomenon is today sufficiently well established that it has become a matter of practical engineering. The electron microscope is designed to operate like an ordinary microscope but using electrons rather than light waves. The much shorter wavelengths available to an electron beam provide greater resolution of detail and enable higher magnifications to be used.

Nor is the wavelike aspect of matter restricted to electrons. De Broglie's idea applies to *all* matter, and the wave properties of alpha particles, nuclei – even whole atoms and molecules – must be taken into account in microscopic physics. Because the wavelength λ diminishes with momentum, macroscopic particles like billiard balls, whose large mass endows them with enormous momentum by subatomic standards, have utterly negligible wavelengths. This is why we do not notice quantum wave phenomena in everyday life.

Equation (2.9) is just of the type which prevents the electron beam from being used to locate a target electron with certainty. To reduce the recoil, slow, low-energy electrons must be sent, and this means low momentum (p) and hence, through (2.9), long wavelengths. But the fact that the electron beam moves as though it were a wave rather than a stream of particles implies that it cannot be used to locate the target electron to within better than one wavelength, so the lower the recoil, the greater the uncertainty in location. We are thus in the same dilemma as with the electromagnetic wave – either the position *or* the momentum may be more or less accurately measured, but not both simultaneously. Similarly, electrons are subject to the other uncertainty relation, (2.8), which prevents their energy being measured to within ΔE unless the measurement occupies at least a time interval $\Delta t > h/\Delta E$. Thus, the uncertainty principle (2.7) is seen to be an *inherent property* of electrons, not merely a result of their interaction with photons.

It is at this stage that puzzlement sets in. The reader may well feel able to accept that electromagnetic waves come in lumps called photons and that these share some of the features of particles, such as carrying energy and momentum. The idea is strange, and no obvious reason has been mentioned why nature should only allow individual quanta of electromagnetic energy to exist, but the situation is easy to visualize. Now we are asked to accept that concrete material particles behave like waves. To make sense of this many people feel tempted to abandon the concept of particles altogether and imagine that matter is really wavelike, with electrons consisting of localized packets of waves bunched together in an approximately coherent particle-like lump. This temptation must be resisted at all costs, for it is quite misleading. One is not asked to accept that an electron *is* a wave, only that its motion is controlled in a wavelike manner.

To clarify what may seem a hopelessly paradoxical situation we shall study in detail a simple experimental arrangement which can actually be used in practice (see Fig. 2.6). A screen contains two small apertures or slits, behind which is placed an electron gun (such as is installed in the back of television tubes) which squirts a spray of these particles towards the slits. Those which line up correctly pass through the apertures and spread out on the other side, rather like commuters emerging from the exit of a railway station. On the remote side of the screen, some distance away, is located a detector which could either be a fluorescent screen like a television, or else a counter device mounted on a track, which collects and registers individual electrons.

Most of the particles are detected in the two regions directly opposite

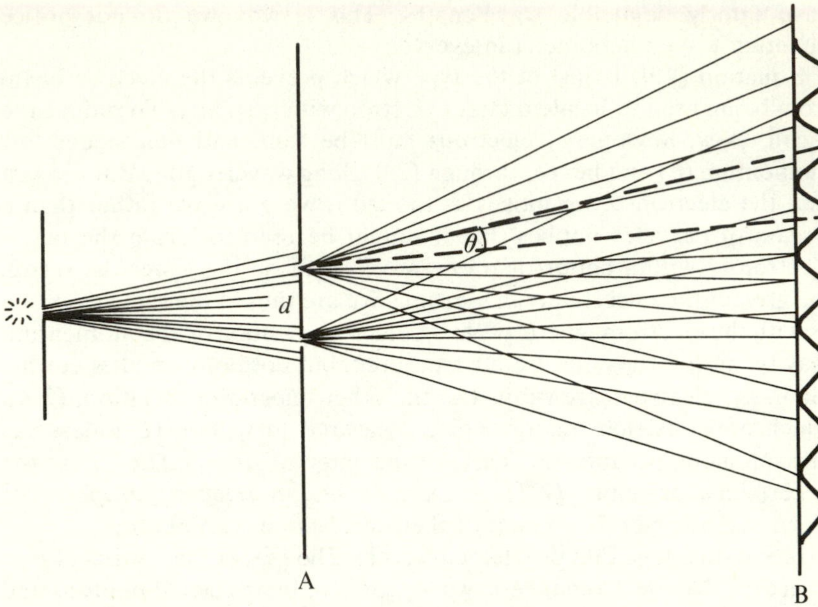

Fig. 2.6. Waves or particles? In this two-slit experiment electrons or photons from the source pass through two nearby apertures in screen A and travel on to strike screen B, where their rate of arrival is monitored. The curve graphed alongside B plots this rate. The pattern of peaks and troughs indicates a wave interference phenomenon. The angle θ between consecutive peak directions is related to the wavelength λ and the slit separation d by the simple formula $\theta = \lambda/d$.

the slits, but between these regions the flux from both slits will overlap. This is where the interference phenomenon occurs most pronouncedly. To understand the significance of it, first imagine that one of the slits is blocked off and the counter is moved along the track behind the screen, collecting the electrons as they come, one by one through the remaining aperture. This way one can measure the average number of particles to arrive in a given time, say one minute, at every position along the track. A graph of this might look something like Fig. 2.7*a*, with a hump centred on a position opposite the slit, and rather less electrons arriving near the extremities of the track. Allowing the particles to come through the other slit instead would produce a similar curve (shown as a broken line) but displaced somewhat to line up behind the other slit. It is most important to emphasize that the electrons are detected as individual particles in this counting process.

Now suppose that we have both apertures open at the same time and collect electrons coming through either of them. We might expect that the result would be to add together the contributions from each slit, yielding a graph of the form shown in Fig. 2.7*b*. This is not so. What is actually observed is shown in Fig. 2.7*c*, and is surprisingly different – a series of humps and dips. In some regions between the slits, where one would expect the contributions from the two apertures to be enhancing one another, there are actually almost no electrons at all arriving at the detector. What strange influence makes the electrons avoid these regions?

This type of pattern – a sequence of humps and dips – is familiar from optics, and a very similar pattern can be obtained by passing not electrons, but *light* though the same system. The phenomenon encountered here is one of *wave interference*, mentioned on p. 49. In fact, it was by using an apparatus of this sort that the physicist and Egyptologist Thomas Young (British, 1773–1829) showed in 1803 that light consists of waves. Are we to believe, then, that electrons are waves? How can this be the case if they can be collected and counted one by one?

It might be imagined that the pattern is produced by some sort of interaction between the electrons (such as an electric force), having the effect of systematically crowding them into certain preferred regions. To investigate this possibility one can turn down the strength of the electron beam, and count for a longer duration instead. As there is less crowding of the particles in the weaker beam, and interaction between the electrons would diminish. Experiment, however, shows that there is no change in the interference pattern. Indeed, it is possible to turn the beam down so far that only *one electron at a time* is traversing the apparatus, thus removing completely any possibility of mutual interaction. Evidently, in some strange way, each electron is interfering with *itself*.

One can go further than this. Suppose we build 10 000 similar pieces of apparatus and just let one electron pass through each. We make a chart and mark a dot at the position that each electron arrives. Continuing in this way we build up a pattern of 10 000 dots, and find that the pattern is identical to Fig. 2.7*c*, with all the peaks and troughs, even though each individual dot represents the result of a *completely separate* experiment occurring at a separate place and time. Nevertheless, the whole assembly of dots considered together shows the same interference pattern as the original, single, apparatus showed, in which there was a steady flux of particles. This remarkable result indicates that the interference phenomenon is really a *statistical* effect and has nothing to do with the collective action between many electrons. It is not that a whole group of electrons influences each other to crowd together in certain regions – for when only one electron passes through each apparatus it can know nothing of the other experiments to be performed somewhere else, of the experimenter's intention to chart the positions of

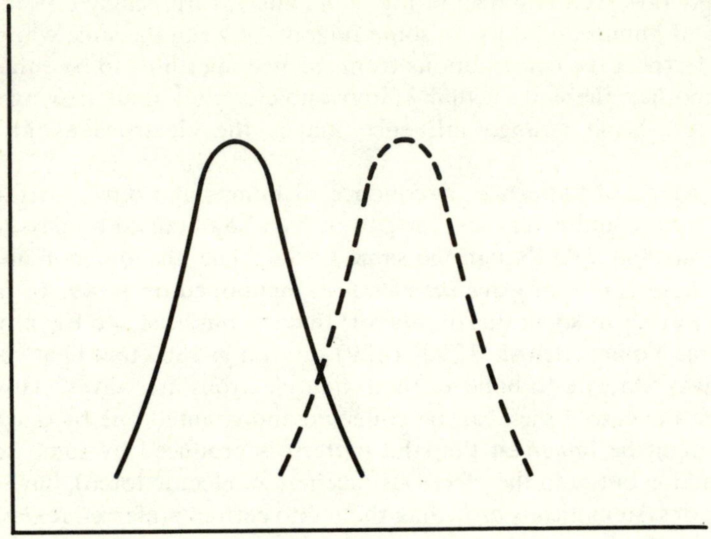

(a)

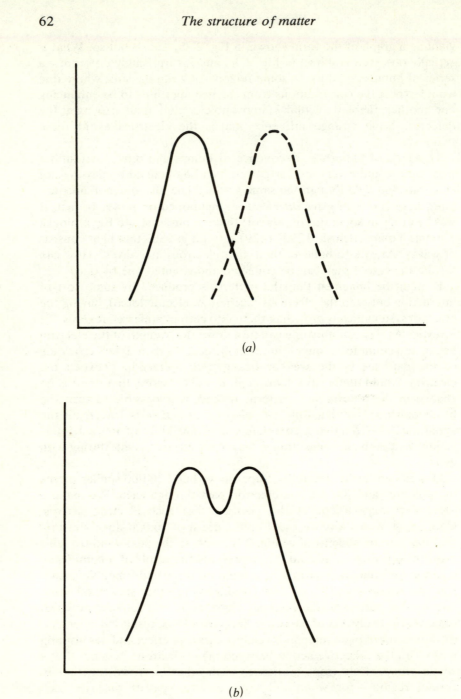

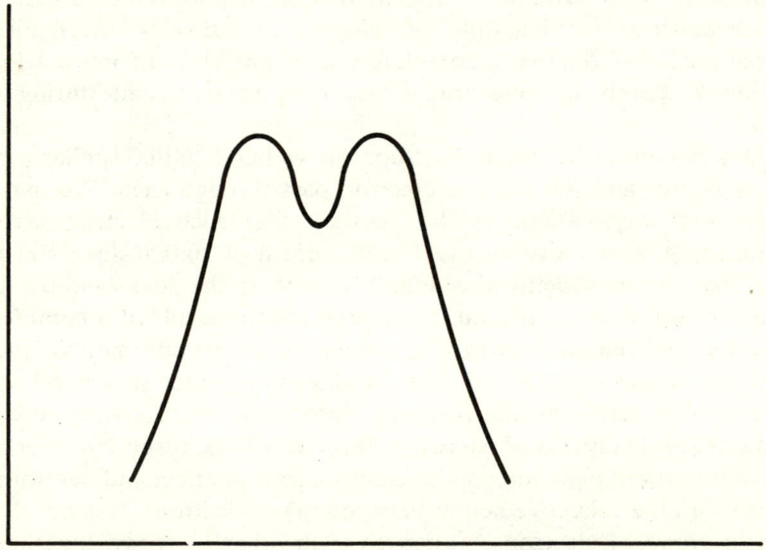

(b)

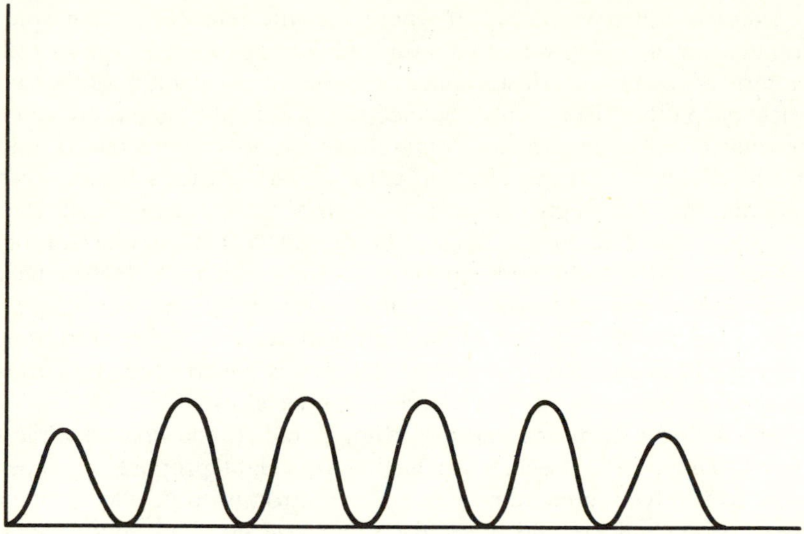

(*c*)

Fig. 2.7. Quantum interference. (*a*) When the right slit of the apparatus shown in Fig. 2.6 is blocked the population of particles arriving (e.g. electrons) peaks behind the left-hand slit. Similarly (broken line) when the left slit is blocked. (*b*) Expected distribution when both slits are open. (*c*) Observed distribution showing peaks and troughs of interference.

all the electrons at the end. The pattern can only be explained by saying that any individual electron is *more likely* to arrive in a region corresponding to one of the high peaks on the graph, than in a trough region. Each electron's arrival is a completely independent event, but because some positionings are more probable than others, more arrivals occur near the peaks of the pattern than the troughs.

The significance of the waves associated with matter is now somewhat clearer. The wave is not a wave *of* matter, or anything else of a tangible physical nature, but a *probability wave*. What is meant by this is that the wave manifests itself though *statistics* by determining the *probability* of the location of a particle. In fact, the amplitude of the wave at a point in space is a measure of the likelihood of finding a particle at that point. Where the wave is strongest, there the particle is most likely to be. In this respect the matter waves are rather like crime waves. A crime wave is not a wave of any object or substance but a measure of the location and frequency of a certain type of event. If a crime wave passes through a town, there is an increased probability that the event of burglary will take place in some particular house. In the same way, if a probability matter wave passes through a region of detectors or counters, there is a high probability that one of them will be triggered.

There is still the mystery of where the interference pattern comes from. After all, when waves pass through an apparatus of this sort, the pattern arises because disturbances from each slit come together and affect each other. But if only one electron is used at a time there can be no question of a disturbance coming through *both* apertures. It is too much to expect that the electron comes apart, diffuses through both slits, and then obligingly reconstitutes itself before entering the detector, wherever that detector happens to be placed. Yet if the electron only passes through one slit, how can the other slit – which the electron never even approaches – influence the final position of the particle, and encourage it to move to one of the high-peak regions of the interference pattern? Remember that if one of the slits is covered up the pattern disappears. Yet only *one* slit at a time can be in use.

This seems a profound and disturbing paradox, and takes a while to digest. Those of a mischievous inclination might propose an experiment to resolve the paradox: when the electron is on its way to one of the slits, act quickly and block the other one up. This action cannot possibly influence the behaviour of the electron passing through the unblocked slit, and yet we know that *two* open slits are needed to obtain an interference pattern. We seem to have a deep contradiction – how can nature escape the dilemma?

Let us examine closely how the mischievous intervention might be accomplished. One way is to shine a strong beam of light (gamma rays would be better) across the entrances to the apertures. We could then spot the approach of an electron by looking for scattered photons, and quickly block off the other aperture when we knew which one the particle is headed for. Recalling the earlier discussion about locating electrons, it is clear that this strategy has a fundamental limitation. The impact of the photons (light beam) with the approaching electron causes a recoil which disturbs its motion. This disturbance is greater the more precise the location of the particle. Obviously we need to locate it to within a region somewhat smaller than the distance between the two apertures, otherwise we could not tell which one it is headed for.

However, according to the uncertainty principle (2.7), this information introduces an uncertainty in the transverse component of the particle's momentum (i.e. parallel to the screen) of about h/d, where d is the slit separation. Now if its forward momentum is p, then the effect of this side-kick of h/d will be to deflect the trajectory through the small angle $(h/d)/p$. But p is related to the wavelength λ of the waves by Equation (2.9), so the deflection angle is

$$\frac{h}{dp} = \frac{\lambda}{d}$$

which is actually independent of h. However, the angle λ/d is precisely the angle between the two peaks of an interference pattern produced by waves of wavelength λ (i.e. the angle θ in Fig. 2.6), known from the theory of optics. Consequently, if we attempt to find which aperture to cover up, the attempt itself destroys the very interference pattern which we are investigating. It is as though nature always contrives to hide her quantum secrets from human cognizance. Even in principle, there is no way that the actual trajectory of the electron can be known and the wave pattern remain intact.

The double-aperture experiment provides a graphic illustration of the wave–particle dichotomy, for both wave- and particle-like aspects of the electrons are manifested in the same experiment. However, they enter in a sort of complementary way. The detection of the electrons has the aspect of particle, but their distribution that of a wave. Notice that we are not forced to make a choice *between* particle and wave; electrons evidently have both wave and particle properties peacefully co-existing. We cannot set one property in conflict with the other because of the inherent uncertainty associated with the Heisenberg principle. Some measurements reveal the particle behaviour of electrons, others the wave behaviour.

The peaceful co-existence is often referred to as wave–particle *duality*, and it brings the properties of the electron much closer to that of the photon. Recall that although light has wave aspects, the photon behaves in many respects like a particle – knocking electrons out of metals, delivering a recoil to a target electron, etc. In this way, electrons and photons can be considered as two types of fundamental particles of nature. In the following chapters we shall see that there are many more. They all share the basic wave–particle duality of the electron and the photon.

The existence of the wave nature of material particles immediately opens up the possibility of resolving the mysteries of the atom. If electrons are subject to wave behaviour, they can no longer be described using Newton's laws, as we did in Section 2.3. If an electron is confined to an atom with a size of about 10^{-10} m, then the uncertainty principle (2.7) tells us that its momentum will be uncertain by at least $h/10^{-10} = 6.6 \times 10^{-24}$ kg m s^{-1}. The electron mass is 9×10^{-31} kg, so the uncertainty in the electron's velocity is roughly $6.6 \times 10^{-24}/9 \times 10^{-31} \simeq 10^{7}$ m s^{-1}, which is greater than the orbital velocity according to the old planetary model using Newton's laws. Moreover, Equation (2.9) shows that at this momentum, the wavelength of the matter waves are comparable to the size of the atom itself. Clearly the wave properties of matter will profoundly modify the atomic structure.

To incorporate these new ideas into physics, it was necessary to replace Newton's laws with some new mechanics consistent with wave–particle duality, the uncertainty principle, the probablistic nature of measurements, and the existence of the quantum of light. This was not an easy thing to do, for Newtonian mechanics had been the foundation of science for over two centuries. The new mechanics was developed mainly by Heisenberg and Schrödinger in the mid 1920s, for which they received Nobel prizes in 1932 and 1933, respectively. Niels Bohr and Max Born (German, 1882–1970, Nobel physics prizewinner, 1954) contributed greatly to its interpretation. It soon gained universal acceptance and acclaim. It is now know as quantum mechanics, and presides over the microcosm. Planck's constant h enters in a fundamental way, and in the limit of large systems the theory reduces to that of Newton. However, its mathematical structure is far removed from that of Newtonian mechanics, and a description of its procedure of application lies far beyond the scope of this book.

The success of quantum mechanics can hardly be overemphasized. Within a very few years the theory had explained not only atomic structure and the features of atomic spectra, but the nature of chemical bonds, interatomic and molecular forces, the structure of solids, the effect of electric and magnetic fields on atoms, the emission and absorption of light and other electromagnetic radiation by atoms, and something about the structure of the atomic nucleus. In more recent years it has explained or predicted the transistor, the laser, superconductivity and superfluidity, the structure and behaviour of liquids, nuclear reactions, the behaviour of subatomic particles, and many other phenomena besides. In addition it has raised profound and as yet unsolved questions about the nature of reality and the place of the observer in the universe. The fundamental limitations on measurement, and the ability for a quantum system to interfere with itself, completely alter the traditional relationship between the human observer and the world he observes. The deterministic certainty of Newtonian mechanics has been replaced by the probabilistic randomness of quantum mechanics and although the manifestations of this underlying uncertainty are mainly restricted to the microworld, nevertheless the issues of principle raised by the quantum revolution have literally cosmic significance.

2.6 *Atomic harmony*

It has been mentioned that the wave properties of matter will have a profound effect on the structure of the atom. Although a proper description of the application of quantum mechanics to atomic structure cannot be given here, a rough idea of the impact of wave–particle duality on this system can be sketched. The properties of ordinary waves are very familiar, and fortunately the probability waves of matter share many of the features of other wave systems, so we may analyse the behaviour of electrons in atoms by appealing to our knowledge of, for example, waves on strings, or membranes.

The problem of the electron bound electrically to the atomic nucleus was the most important of all applications of the new mechanics, in view of the serious inconsistencies between classical electrodynamics and what is known of atomic structure and spectra: in particular, the existence of favoured energy levels in which the electron remains stable and non-radiative. The confinement of an electron to an atom is rather like that of a particle shut inside an invisible box. The wall of the box is the electric field of the nucleus, which tries to pull the electron back from the periphery of the atom. Of course, this is not a rigid, impenetrable wall, but in its usual condition in the atom the electron has insufficient energy to penetrate this rather flabby barrier, so it is trapped by the nucleus as effectively as if there were a wall.

According to the ideas of quantum mechanics, the situation that we have to study is that of a *wave* confined inside a flabby-sided box. The full treatment of this is not hard, but does involve some undergraduate level mathematics. Fortunately, however, most of the qualitative features can be deduced by studying a simplified model – a particle (or wave) moving *freely* inside a *rigid* box. This means that, for the purposes of illustration, we assume that the electron is unaffected by the electric field of the nucleus until it encounters the periphery of the atom, when it feels a sharp barrier. Physically, then, the problem is reduced to that of a freely propagating wave inside a rigid sphere. A very similar, but slightly more familiar arrangment, is the one-dimensional analogue: wave motion confined to a stretched string with fixed ends – a guitar string, for example (see Fig. 2.8). In this system, the vibration of the string represents the quantum wave and the fixed ends of the string reflect back the waves in the same way that the walls of the box reflect back the electron. It turns out that the behaviour of a plucked guitar string is remarkably similar to the behaviour of an electron inside an atom.

Everyone is familiar with the phenomenon of waves travelling along a string, which can easily be produced by wiggling a free end up and down. If the remote end of the string is held fixed, then the waves will reflect towards the free end again. If both ends are held fixed and the centre of the string is disturbed, the waves reflect back off both ends and the total motion can be a complicated pattern. However, there are certain natural, simple motions that can occur on a stretched string with fixed ends which involve standing (or stationary) waves, rather than travelling waves. This pattern of disturbance is a coherent vibration in which all parts of the string move up and down in synchronism at a common frequency.

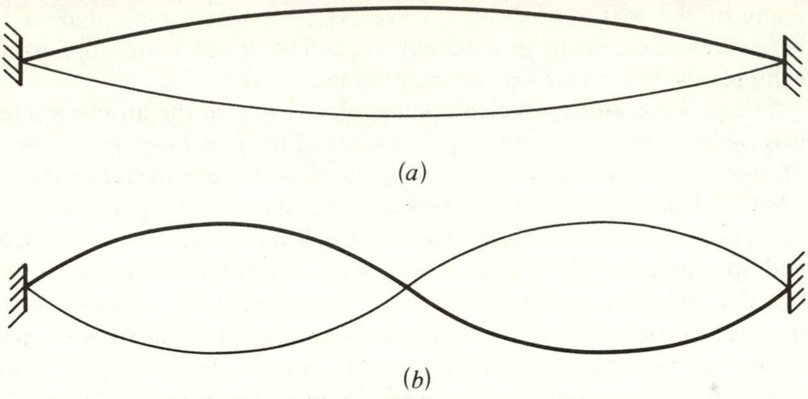

Fig. 2.8. Standing waves. A vibrating guitar string traces out a simple pattern. (*a*) The fundamental (lowest) note is obtained when the shape is as shown. (*b*) This shape gives a note one octave higher than (*a*). Plucking the string at random will generate a superposition of both wave shapes, and higher harmonics too.

The simplest type of synchronous motion is shown in Fig. 2.8*a* where the centre of the string suffers the maximum displacement. This type of motion will usually be excited by plucking the guitar string at the centre and will produce a note of fixed frequency depending on the length of the string, its density and its tension. Another possibility is shown in (*b*) where the centre of the string remains undisplaced and is a point of symmetry. The wavelength is exactly one half that of (*a*), so according to the relation (1.5) the frequency will be twice as high. A whole hierarchy of vibrations can occur with appropriate plucking, each with a frequency which is an exact multiple of the lowest one. Notice the important point that not *any* frequency can occur. The reason for this is that only waves in which an exact number of (half) wavelengths fit on the string are allowed, otherwise the string would not remain still at the ends.

Recall now the characteristic feature of atomic spectra. The light waves emitted or absorbed by atoms do not have any frequency, but are restricted to certain favoured ones. Physicists soon found that the favoured frequencies were not random but related by simple numerical formulae. A glance at Fig. 2.4 shows the systematic arrangement of some spectral line frequencies. If the frequency of light is related to the motions of the atomic electrons we would expect the frequencies of the matter probability waves inside the atom to satisfy simple numerical relationships too. This suggests that the electrons reside in the atom in much the same way as a synchronous wave resides on a guitar string – in a standing or stationary wave pattern.

The stationary states in which the electron matter wave is trapped around the nucleus can easily be calculated. This gives both the shape of the wave and its frequency. As with the guitar string not every frequency is allowed; only those that fit an exact number of wavelengths into the spherical 'box'. The frequency of the matter waves is related to the *energy* of the electron through the fundamental Planck formula (2.5) so we arrive at the important result that the electrons must reside in certain definite, preferred, *energy levels*, corresponding to the standing waves. Moreover, these levels enjoy the same simple numerical relationships as do the wave frequencies (see Fig. 2.9). The calculation shows that the energy levels are given by the simple formula $-2\pi me^4/n^2h^2$ where $n = 1,2,3, \ldots$ (the energy is negative because the electrons are bound). This formula neglects certain small effects due to relativity, the interaction of the electron with its own magnetic field and that of the nucleus, and other subtle corrections.

It is now possible to explain two important things at once. Firstly the discrete nature of atomic spectra: when an electron is in one of the energy levels, it is in a stationary state. It cannot radiate any amount of energy in an electromagnetic wave as it circles the nucleus because it would then have to reduce its energy by a corresponding amount. But only certain values of the energy – those corresponding to the energy levels – are allowed. Hence it may only radiate by emitting a large packet of energy which would transfer it right down to one of the lower lying energy levels. The atom thus emits a quantum of light (photon) with a particular energy corresponding to the energy gap between the levels. According to (2.5) this will be a photon with a definite, fixed frequency. Similar photons of that frequency will be emitted by other surrounding atoms.

The existence of a *series* of spectral lines is also easily explained. An electron may arrive at any particular level (e.g. G in Fig. 2.9) by making a downward transition from a whole sequence of excited levels 1,2,3, . . . Quantum mechanics can also be used to calculate the relative *probabilities* of these different transitions and thus predict the relative *intensities* of the corresponding spectral lines: highly probable transitions will occur in many atoms and produce a lot of light at that frequency.

The second fact that can be explained is the stability of the atom. Recall that in the classical picture an electron is free to spiral closer and closer to the nucleus, radiating without limit. In the quantum picture this is not allowed. Just as there is a lowest frequency note on a guitar string (the one is shown in Fig. 2.8*a*) so there is a lowest energy stationary wave around the atom. This energy level is called the ground state, and the ones lying above it are called excited states. The existence of a ground state of *lowest* energy is an immediate consequence of the

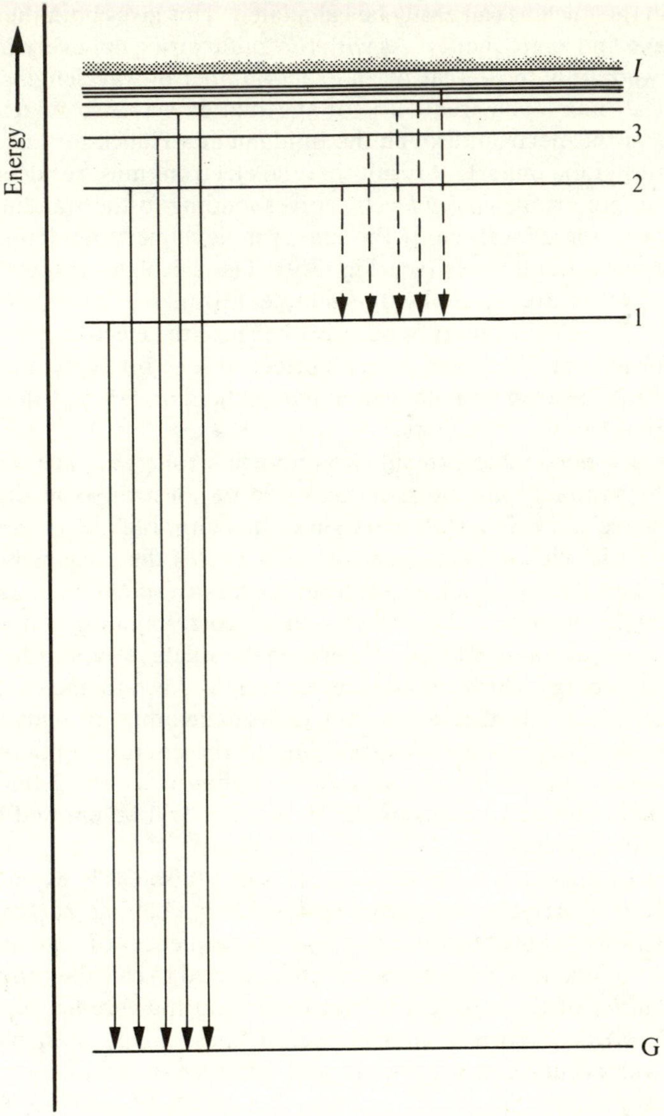

Fig. 2.9. Atomic energy levels. The orbital levels of the hydrogen atom form a simple series in energy. G is the ground (lowest energy) level, above which the first, second, third, . . . excited levels (labelled 1, 2, 3, . . .) converge onto *I*, the energy at which the atom becomes ionized and the electron detaches itself. Above this level the electron may move freely with any energy (shaded region). The arrows indicate electron transitions between levels, which result in the emission of a photon. The simple sequence of levels is reflected in the simple pattern of lines shown in the hydrogen spectrum (Fig. 2.4). Each level has an associated series of transitions (see arrows): only two series are shown.

uncertainty principle. As already remarked, to lower its energy an electron has to approach closer to the nucleus, but to confine a particle near to the nucleus induces a large uncertainty in its momentum which means that it is likely to be moving very fast. When the electron is sufficiently close to the nucleus the energy associated with this motion offsets the electric energy gained by drawing closer to the attracting nucleus. The total energy cannot then be further reduced. This corresponds to the ground state. Crudely speaking, the quantum uncertainty provides a type of repulsive buoyancy to prevent the electrons from collapsing on to the nucleus.

In its normal condition an atom will be in its ground state with all electrons having the minimum allowed energies. This state is stable. It is possible for the atom to become excited, for example if irradiated with photons, and for one or more electrons to make an upward transition to a higher energy level. Shortly afterwards it will return to the ground state again with the emission of a photon. Quantum mechanics enables one to calculate how long the processes of excitation and decay are likely to take, i.e. the expected lifetime of the initial state. Most excited states live only for a tiny fraction of a second.

In a real atom the number of allowed energy levels is unlimited, but they crowd closer together at the higher levels (see Fig. 2.9). If an electron receives sufficient energy it may jump over all these bound states entirely and escape from the atom, a process called ionization. It is a simple matter to calculate the energy needed to ionize, say, a hydrogen atom from its ground state, and the theory coincides perfectly with the measured value.

In Fig. 2.10 we have plotted the probability wave profile for the ground state of the hydrogen atom. Notice that nowhere is it zero (except at the exact centre of the nucleus). Remember that this profile represents the probability of finding the electron at that particular radius. We see that in the quantum theory picture the electron does not orbit the nucleus at a certain fixed radius, as does the Earth around the sun. There is a definite probability of finding the electron right up against the nucleus, or far away. The curve is fairly strongly peaked around the most probable position which, according to the wave mechanics calculation referred to above, lies a distance $h^2/4\pi^2me^2 \simeq 0.53 \times 10^{-10}$ m from the centre, and falls very rapidly to zero at large radius. Of course, in principle it would be *possible* to find the electron as far away as the moon, but the probability is absurdly small.

The inherent uncertainty in the electron's radial position which this illustrates is also true of its angular location. It is not possible to discuss the orientation of the electron relative to the nucleus, only the probability of its being in a certain region. In the ground state it is equally

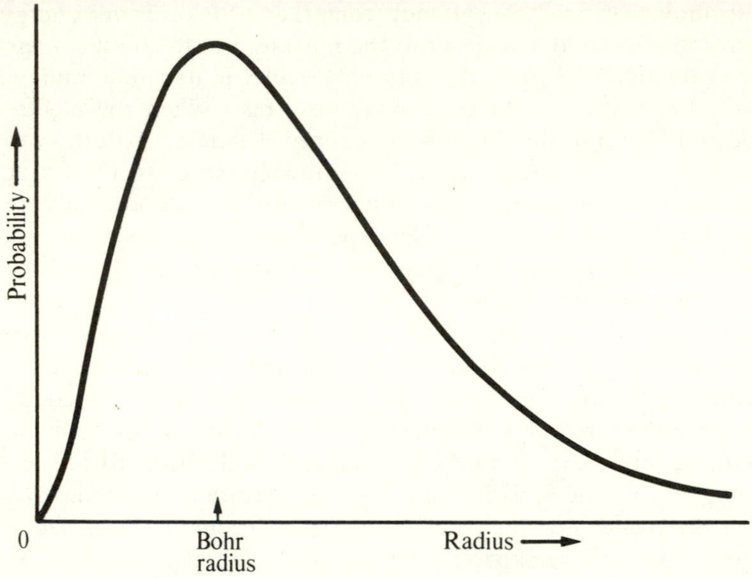

Fig. 2.10. Probability wave for an atomic electron. This graph shows a radial section for a hydrogen atom in its ground state, corrected for spherical geometry, so that the profile indicates the relative probability of finding the electron at any given radius. Although peaked around one Bohr radius ($h^2/4\pi^2me^2$), there is a finite probability of finding the electron close to the nucleus (near radius = 0), or very far away.

probable for it to be found at any orientation around the nucleus, so it is not even possible to say whether it is moving clockwise or anticlockwise. Indeed, it turns out that the angular momentum (a measure of its inertia of rotation) is actually zero for this state, something which is a total impossibility in Newtonian mechanics. This means that it doesn't really make much sense to think of the electron as rotating around the nucleus at all. In the ground state we can envisage the electron as no longer experiencing any centrifugal force. It is prevented from falling on to the nucleus solely by the quantum uncertainty energy discussed on p. 71, which buoys it up against the electric attraction.

One important feature of atoms has yet to be explained. The picture so far is of a nucleus surrounded by a cloud of electrons stacked up in a sequence of definite energy levels. The low-lying (least energy) levels are occupied by electrons orbiting close to the nucleus, the higher levels are farther out. We may think of this as a sequence of *shells* sitting inside each other. As explained above, an electron residing in one of the outer levels, or shells, is unstable because it can rapidly emit a photon and jump down to one of the lower energy levels. This suggests that the electron cloud will collapse like a house of cards on to the lowest available level. The evidence of chemistry and X-ray scattering shows that the electrons are not all in the same shell, but are distributed over a wide range of shells.

The explanation of this atomic rigidity requires a further development of the ideas of quantum mechanics into the domain of relativity.

2.7 *Quantum theory and relativity*

Planck's basic quantum hypothesis, Equation (2.5), written in terms of photon wavelength λ is

$$E = hc/\lambda.$$

On the other hand, matter waves are found to have a wavelength determined by their momentum p through the relation (2.9)

$$p = h/\lambda.$$

If this equation is taken to apply to photons as well as material particles we may eliminate λ between these two equations to obtain

$$E = pc \tag{2.10}$$

which is precisely Equation (1.12) obtained from relativity theory. This confirms that the quantum mechanics of the photon is consistent with the principles of relativity, provided the photon is regarded as having a zero rest mass. For that reason it is sometimes called a 'massless' particle, but it is important to remember that only *rest* mass is being referred to. For a particle with non-zero rest mass, Equations (2.5) and (2.9) are no longer compatible because (2.10) must be replaced by the more general relation (1.10)

$$E^2 = m_{\text{rest}}^2 c^4 + p^2c^2.$$

In the late 1920s attempts were made to modify the theory of quantum mechanics for material particles to make it compatible with the concepts and spirit of the theory of relativity in general, and Equation (1.10) in particular. The starting point was the equation which had been discovered by Schrödinger in 1926 that describes the propagation of matter waves. This equation suffers from the same defect as Newtonian mechanics: it does not transform correctly when the reference frame is changed. As in Newtonian mechanics, so in quantum mechanics, this defect is of little consequence when the bodies of interest are moving slowly, but we have already seen that inside atoms electrons are moving at an appreciable fraction of the speed of light so the motivation for finding a relativistic version of quantum mechanics is even greater.

Mathematically the defect appears as an asymmetry between the way in which space and time enter into the Schrödinger wave equation. Correcting this asymmetry, and placing space and time on an equal footing, is equivalent to constructing a wave equation with the correct relativity transformation properties, which is, moreover, consistent with Equation (1.10). This can be done in two ways: by changing the time part to look like the space part, or vice versa. The first modification is the easiest and leads to an equation for a wave that is very similar to that for an electromagnetic wave, except that there is a rest mass term. The second method is more difficult and involves introducing an entirely new mathematical object, called a spinor, into the modified space part of the equation. The reason for this modification is connected with geometry and the fact that space has three dimensions while time has only one. The resulting equation, discovered by the British physicist Paul Adrien Maurice Dirac in 1928, is of an entirely new type and immediately produced some surprises.

In a sense Dirac's equation is like the 'square root' of the usual wave equation, and so it throws up not Equation (1.10) but its square root

$$E = \sqrt{(m_{\text{rest}}^2 c^4 + p^2c^2)}. \tag{2.11}$$

However, every quantity possesses *two* square roots, one being the negative of the other. Thus the equation,

$$E = -\sqrt{(m_{\text{rest}}^2 c^4 + p^2c^2)} \tag{2.12}$$

in which E is *negative* is equally compatible with Equation (1.10), and inevitably appears among the solutions to Dirac's equation. It would seem that Dirac has produced an equation which describes *negative energy matter*, as well as ordinary matter. The significance of this curious bonus will be explained in Section 3.3.

In addition to the negative energy solutions, Dirac found another completely novel feature about his solutions. Even when the particle which they describe is at rest, it still behaves as if it is in some sense rotating, so it must represent an *instrinsic spin* of the particle, a sort of internal angular momentum.

Superficially it might seem as though intrinsic spin can be envisaged as a rigid rotation about an axis of an extended body – a little spinning sphere perhaps, like a miniscule Earth, turning about a line through its poles. Closer examination, however, reveals that things are not so simple. Indeed, it is probable that instrinsic spin has no counterpart in familiar objects at all, but is an inherently quantum phenomenon. For example, this spin behaves in a very strange way under rotations. To understand this, suppose that the spinning Earth were tipped over end to end so that the north pole was located where the south pole is now. Then its direction of rotation would clearly change from, say, clockwise to anti-clockwise as far as a fixed observer (e.g. one on the moon) is concerned. A further turn of 180° would bring everything back to the way it started, with the spin pointing the same way and rotating in the same direction. Now do the same thing to a Dirac particle. What is found is that after a full rotation of 360° the particle is *not* returned to its original state. Only after *two* full rotations does it return. This enigmatic geometrical property is a consequence of the appearance of spinors in Dirac's wave equation.

A further odd property of spin, which is also true more generally of rotation in quantum mechanics, is that the spin 'axis' cannot point in any direction. If an experiment is set up to measure how many particles have their spin pointing along a particular direction then it is found that the spin must either point 'up' or 'down' this direction, but nowhere in between. The chosen direction can be varied at will, but whenever a measurement is made, only one of two alternatives can result. Thus, Dirac particles have discrete spin levels in much the same way as they have energy levels.

Another fact to emerge from the Dirac equation is that just as the direction of spin is restricted, so is its quantity. In units of angular momentum, a Dirac particle must have an amount of spin equal to $h/4\pi$, so once again Planck's constant makes its appearance. For arithmetical convenience, physicists usually use the constant $h/2\pi$ rather than h, the former being denoted by $\hbar$. In this unit, the Dirac particle has spin one half (the $\hbar$ being understood). We shall often use this terminology in the following sections.

If the particle carries an electric charge then it seems likely that its rotation will endow it with an intrinsic magnetic field, because in a crude sense the spinning charge represents a circulating current (even though the particle is not really an extended body in rigid rotation, but more like a point mass). The equations show that this is indeed so, but because the spin is 'peculiar' so is the magnetic field – it is *twice* the value it would have been had the electron really been a little rigid, rotating ball. This double value is closely related to the double-valued nature of the spin under rotations through 360°, and is further evidence that intrinsic spin has no real classical counterpart.

Following this mathematical analysis it soon became clear that Dirac's equation correctly describes the behaviour of *electrons*. The evidence of atomic spectra had long suggested that some sort of intrinsic spin was at work, which caused small effects that Schrödinger's wave equation could not adequately explain. For example, when an electron orbits in an atom it sets up a magnetic field. This field will interact with the magnetic field caused by the intrinsic spin; that is, the electron behaves like a tiny bar magnet and tries to flip around in the magnetic field caused by its orbital motion. However, its direction must be either parallel or antiparallel to the field. There is an energy bonus in flipping antiparallel to the field because north magnetic poles are attracted to south magnetic poles. It follows that there is a tiny, but measurable, energy gap between spin-up and spin-down electrons. An electron can change from spin up to spin down if it emits a very low-energy photon to take up the energy difference, but for many processes the excited electron will drop to a lower level before this happens. The energy of the emitted photon will then depend on which spin direction the electron had in the excited level. The orbital transition can thus produce photons with either of two slightly different frequencies. This difference shows up in atomic spectra as a *splitting* of spectral lines into doublets, an effect known as fine structure. Fine structure is a direct confirmation of the existence of intrinsic spin (see Fig. 2.11).

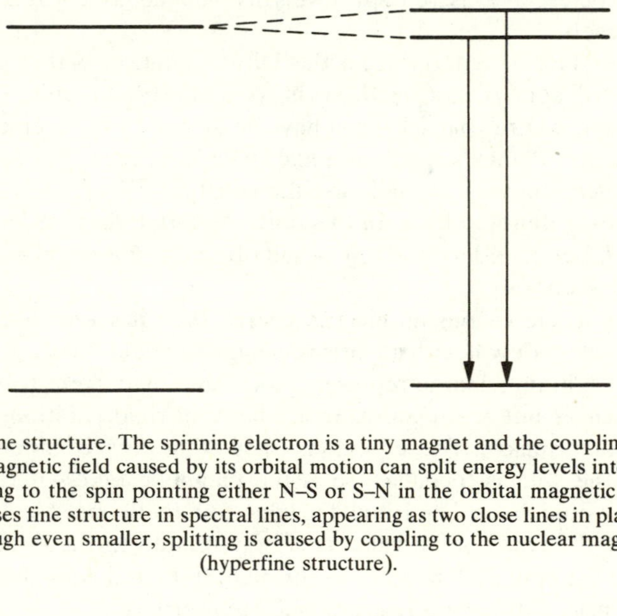

Fig. 2.11. Fine structure. The spinning electron is a tiny magnet and the coupling between it and the magnetic field caused by its orbital motion can split energy levels into doublets corresponding to the spin pointing either N–S or S–N in the orbital magnetic field. The splitting causes fine structure in spectral lines, appearing as two close lines in place of one. Similar, though even smaller, splitting is caused by coupling to the nuclear magnetic field (hyperfine structure).

Using the Dirac equation, physicists were able to recalculate the energy levels of atoms, including the effects of relativity. Not only fine structure and spin phenomena, but also effects such as the relativistic variation in the electron's mass with its velocity as it orbits the nucleus are taken into account. In simple atoms, such as hydrogen, the accuracy of Dirac's equation is truly phenomenal. Even the tiny effect caused by the coupling of the magnetic field of the nucleus to the electron spin can be correctly calculated and accurately measured. However, there remain discrepancies in the measured and calculated energy levels that are so tiny that they represent a fraction as small as 10^{-7} of the electron's binding energy. After the second world war, these tiny discrepancies were explained and calculated correctly using the quantum theory of fields. This subject will be taken up in Section 4.1.

For his work on atomic theory, Dirac shared with Schrödinger the 1933 Nobel physics prize.

It might be wondered whether we could assemble billions of electrons together with their spins all pointing the same way and moving in a coherent fashion. This would enable us to create something like a macroscopic fluid which continued to display the peculiar phenomena associated with Dirac particles. In 1925 the German physicist Wolfgang Pauli (1900–58) proposed that there is a fundamental principle which prevents Dirac particles from behaving in concert to make a macroscopic matter wave. Called the exclusion principle, it states that no two Dirac particles are allowed to occupy the same quantum state. The principle has many far-reaching consequences ranging from the structure of atoms and their nuclei, to the gravitational collapse of stars, and earned Pauli a Nobel prize in 1945.

One immediate application of the Pauli principle is to the organization of electrons in the energy levels of a heavy atom, a topic discussed in the previous sections. It has been explained how electrons in excited levels are unstable against emitting a photon and dropping down to the ground state. This raises the mystery of why a normal atom does not have all its electrons crowded into the lowest level. However, according to the Pauli principle, apparently only one electron is allowed down there. Taking into account the two spin directions, there is in fact a maximum of *two* electrons in the lowest orbital level. So long as this level is filled, another electron has to go in the next one up. It is quite stable there as it is forbidden to make a downward transition. Repeating the reasoning, one sees that the electrons have to stack up on top of each other in a sequence of larger and larger shells, exactly as described on p. 72.

Another important application of the Pauli exclusion principle is to the statistical behaviour of a large collection of electrons. Inside a lump of metal, there are perhaps 10^{24} electrons roaming about more or less freely, for reasons explained at the end of Section 2.2. These electrons act rather like a gas, and their combined behaviour can be analysed using an averaging procedure. If the electrons were ordinary classical particles then the distribution of energy partitioned among them would look something like Fig. 2.12*a*, assuming that they are in equilibrium at some temperature. However, they are not 'ordinary' for two reasons. The first is the fact that electrons, unlike ordinary particles, must be

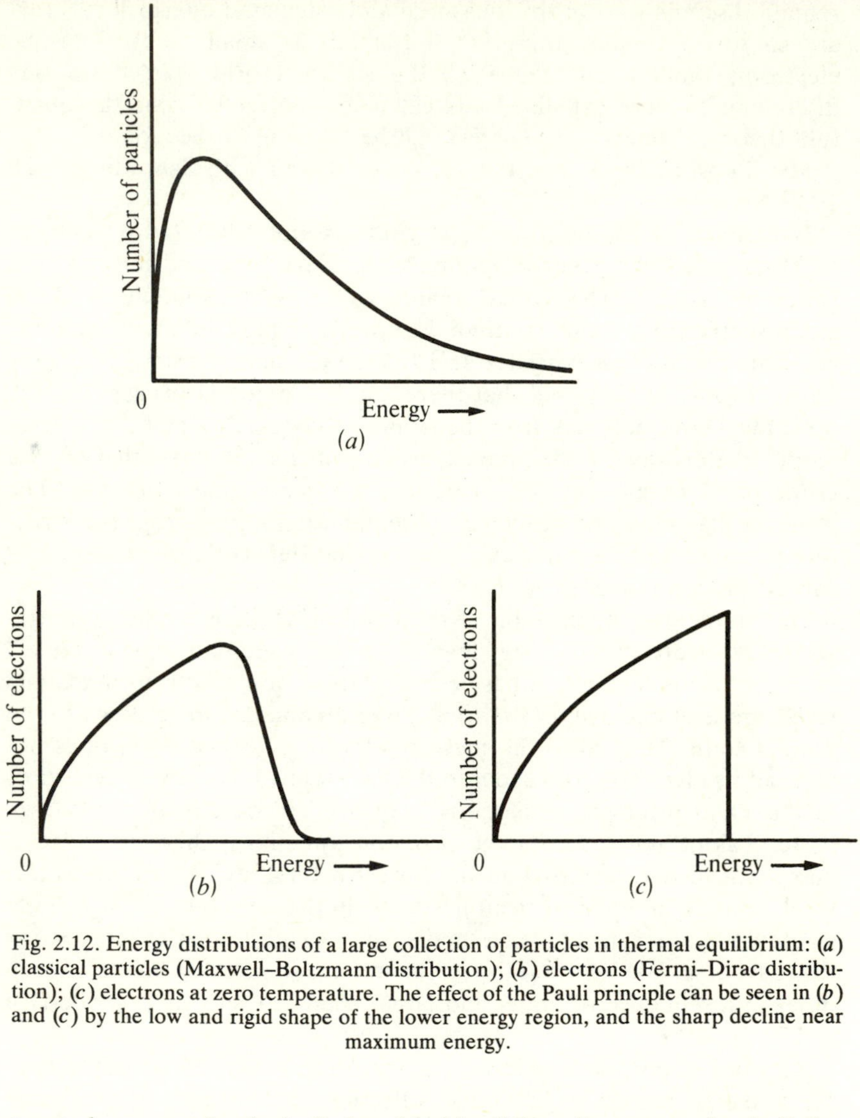

Fig. 2.12. Energy distributions of a large collection of particles in thermal equilibrium: (*a*) classical particles (Maxwell–Boltzmann distribution); (*b*) electrons (Fermi–Dirac distribution); (*c*) electrons at zero temperature. The effect of the Pauli principle can be seen in (*b*) and (*c*) by the low and rigid shape of the lower energy region, and the sharp decline near maximum energy.

treated as completely indistinguishable. When far apart, the indistinguishability has little consequence, but when close together it is very important because of the uncertainty associated with their position. We can never determine which electron is where. This must be allowed for in the crowded conditions inside a metal.

The second reason is the Pauli exclusion principle. There is a natural tendency for most electrons to acquire an energy close to the average value, a sort of equal sharing principle. It is an effect which would show up in the sharply peaked nature of Fig. 2.12*a*. Instead, because of the Pauli principle, the lower and average levels fill completely, obliging the remaining electrons to maintain higher than average energy. The distribution can be calculated to look something like Fig. 2.12*b*.

The shape of this graph has important consequences for the behaviour of metals. If the metal were cooled, even to absolute zero, the electrons would still be obliged to retain a large amount of energy. The distribution would look like Fig. 2.12*c*, with all the levels up to some maximum energy completely filled and all the higher levels empty. This is the lowest energy state that can be achieved. As heat is added so some of the electrons near the top levels can absorb energy and make transitions upwards to some of the empty levels there. However, this upward mobility is extremely restricted by the fact that only the electrons near the top levels can move. The lower ones cannot get enough energy from the heat to raise themselves up over the whole range of occupied levels above them. The outcome of this rigidity is that 'boiling' electrons out of a metal is a very inefficient process. So strong is the rigidity that when the sun burns out and it shrinks to a size comparable to that of the Earth, even its immense weight will be supported against further gravitational collapse by the Pauli exclusion principle.

At this stage we should recall that Dirac's equation was only one of two alternative relativistic modifications of Schrödinger's wave equation. The other is known as the Klein–Gordon equation, after the physicists Oskar Benjamin Klein (Swedish, 1894–1977) and Walter Gordon (German, 1893–1939). Does it, too, have a role to play in describing subatomic particles? The Klein–Gordon equation does not contain solutions which describe intrinsic spin. In the coming sections we shall see that there are many different kinds of subatomic particles other than electrons. Many of these also have spin one half and are correctly described by the Dirac equation, but not all. Some are known which have no spin, and the Klein–Gordon equation applies to them.

If we think of the photon as a particle then it can be shown from Maxwell's equations to carry a spin of one, or twice that of the electron. Moreover, we have to face the fact that if quantum mechanics is a fundamental feature of nature then it must apply to all the forces of nature, including gravity. Otherwise we could (in principle) use gravitational effects to violate the Heisenberg principle. The subject of quantum gravity is still in an unsatisfactory state, and will be discussed further in Section 6.4, but for now let us suppose that there exist *gravitons* – quanta of gravitational waves – just as there exist photons. The reason we do not notice gravitons is because gravity is so weak. These gravitons can be thought of as massless particles moving at the speed of light and obeying Equations (1.12) and (2.5). In addition it can be shown that they possess an intrinsic spin of two, which is twice that of the photon and four times that of the electron.

As we shall see, the evidence of subatomic particle physics indicates that we must be prepared for particles with *any* integral multiple of $\frac{1}{2}\hbar$, as well as spinless ones. The rotation properties of the Klein–Gordon, Maxwell and gravitational field equations do not display the 'peculiar' properties associated with Dirac particles. These are the integer spin particles, having an intrinsic spin equal to zero or a whole number of units of $\hbar$. In contrast, particles with spin one half, three halves and, in general, *half*-integral units of $\hbar$, do share the spinor rotation peculiarities. This suggests that we can divide subatomic particles into two classes: those, such as the photon and graviton, that have integral spin, and the rest, such as the electron, that have half-integral spin. For historical reasons the former are called *bosons* after the Indian physicist Satyendra Nath Bose (1894–1974) and the latter are called *fermions* after the Italian physicist Enrico Fermi (1901–54, winner of the 1938 Nobel physics prize).

The significance of the division is that the Pauli principle only applies to *fermions*. So a collection of photons or gravitons can be crowded together and used to build up coherent, macroscopic wavelike behaviour. It is precisely this reason why electromagnetism and gravity were long known as classical macroscopic fields, whereas the matter waves associated with fermions (e.g. electrons) do not manifest themselves on a macroscopic scale. Suppose we wished to build up a macroscopic matter wave of electrons at a frequency of, say, 1 MHz. According to quantum theory, the frequency ν is related to the energy E of the wave by the Planck formula $E = h\nu$, so to produce a macroscopic effect it is necessary to superimpose billions of electrons with the *same* energy E, and hence frequency ν. But the Pauli principle permits at most *two* such particles to have the same energy. We see that fermion waves are forever destined to remain microscopic in relevance. That is why they were not discovered until the 1920s.

2.8 *The constants of nature*

The laws of nature which have been discussed in the previous sections have consisted of two elements. First, they describe relations between variable quantities such as force, distance, time, electric field and so on, and secondly they contain numerical constants such as G, c and h. The constants determine how large or small the actual measured values of the variable quantities will be compared to some standard length, or interval of time, for example.

As an illustration consider Newton's law of gravitation described by Equation (1.3)

$$F = \frac{GM_1M_2}{r^2}$$

This universal relation tells us that given two masses of M_1 and M_2 a distance r apart, a certain gravitational force F is exerted between them. The F, M_1, M_2, and r quantities may vary considerably, and the equation describes how the variation of one of them affects the others. But G is a universal constant of nature, a fixed number. If the masses are in kilograms, r is in metres and F is measured in Newtons, then G is found experimentally to be the number 6.67×10^{-11}N m^2 kg^{-2}. So long as we use these units G is *always* that number.

Let us make a list of all the constants of this type that have been discussed so far. The analogues of G for electricity and magnetism were mentioned briefly in Section 1.3, but as pointed out on p. 23 these two quantities are related to the speed of light so we shall just add c to the list. The quantity c also has the important property that it relates mass to energy through the formula $E = mc^2$. In Section 2.3 it was described how Planck discovered a new constant h which has a most profound significance. When the atomic theory of matter became well understood, it was realized that nature provides us with certain other natural numbers that are apparently the same throughout the universe. One very basic number is the fundamental unit of electric charge, e, which occurs on the electron and hydrogen nucleus. Another is the mass of the electron, m, and of the hydrogen nucleus, M.

For a long time physicists have been intrigued as to the significance of these numbers, why they have the values they do and not something else. Some theories have been proposed which attempt to link them together in a grand scheme, though none of these theories has been widely accepted. To achieve such an objective, the values of the constants cannot simply be taken as they stand. The reason for this is that G, c, h, e, m and M share the property that their numerical values depend on the units used to measure them. For example, the speed of light is either 3.0×10^8 or 6.7×10^8, depending on whether we use metres per second or miles per hour. Nature cannot know the units we wish to use, so any constant which has significance must be expressed as a *ratio* of quantities with like units. One simple example concerns the masses M and m. If we take the ratio M/m we obtain the number 1836.152 whether we use kilograms, ounces or tonnes to scale M and m individually. This is sometimes expressed by saying that M/m is dimensionless (i.e. has no units, is a pure number).

Several interesting dimensionless ratios may be made out of our constants. First, there is the famous ratio

$$\frac{e^2}{\hbar c} = \frac{1}{137.036} \tag{2.13}$$

known as the fine structure constant, for reasons which will be explained below. A second is

$$\frac{GMm}{e^2} \simeq 10^{-40}. \tag{2.14}$$

This almost unimaginably small number is the ratio of gravitational to electric forces between the two constituents of a hydrogen atom, as may be seen by eliminating r between Equations (1.3) and (1.4). Finally the number

$$\left(\frac{G\hbar}{c^3}\right)^{\frac{1}{2}} \simeq 10^{-35}\ \mathrm{m} \tag{2.15}$$

is a curious one. It has the units of length and so is not a dimensionless ratio. However, another number with units of length is

$$\frac{\hbar}{Mc} \simeq 10^{-15}\ \mathrm{m} \tag{2.16}$$

so that the ratio of these two lengths is about 10^{-20}. (This relation is close to the square root of ratio (2.14).)

The physical significance of these numbers can be guessed by examining the quantities contained in the ratio. For example, $e^2/\hbar c$ combines features of electrodynamics (e^2), quantum theory ($\hbar$) and relativity (c). It might therefore be expected that 1/137.036 would occur prominently in relativistic quantum electrodynamics, and this is indeed the case. In Section 2.7 it was explained that relativity and quantum mechanics combine together to predict that an electron possesses an intrinsic spin and, because of its charge e, an intrinsic magnetic field. This field couples to the electron's orbital magnetic field inside the atom and has the effect of splitting the energy levels into doublets. The evidence for this comes from atomic spectra, where a close examination reveals that some lines which at first sight appear single, are in reality split in two, a feature known as fine structure (see p. 76). The amount of splitting is determined, as expected, by the ratio $e^2/\hbar c$, and it is for this reason that it is called the fine structure constant. In fact, this ratio has a much more general significance; it determines the strength by which all charged subatomic particles couple to the electromagnetic field, so among other things it fixes the rate at which electrons emit photons when accelerated. In essence, it gauges the *importance* of quantum electromagnetic processes.

As another example, consider the length unit $(G\hbar/c^3)^{\frac{1}{2}} \simeq 10^{-35}$ m. This length is the smallest natural unit of physical significance known. It combines features of gravity (G), relativity (c) and quantum theory ($\hbar$) and might therefore be expected to have something to do with quantum gravity. In analogy with the role of $e^2/\hbar c$ one imagines that $(G\hbar/c^3)^{\frac{1}{2}}$ determines such things as the rate at which gravitons are emitted by accelerated masses. If these expectations are correct, then it suggests that the effects of quantum gravity (e.g. graviton processes) will only be important on an unimaginably tiny scale, when the dimensions of the system of interest are less than 10^{-35} m. This is a full 20 powers of 10 less than the size of an atomic nucleus, and seems to indicate that quantum gravity effects can always be ignored for practical purposes. It also suggests that the graviton will never have a significant role to play as a subatomic particle, unlike its counterpart, the photon. Unfortunately, fundamental difficulties with the theory of quantum gravity (see also Section 6.4) preclude a definite statement about this.

3

Breaking matter apart

Rutherford's analysis of the scattering of alpha rays from thin foils established the basic morphology of the atom, particularly the existence of a highly compact nucleus in which most of the matter resides. Neither theorists nor experimenters could be expected to accept the existence of this central body without wondering about its internal structure. What is it made of? Why are there so many different kinds of nuclei? Why do they always have a positive electric charge equal to a whole number multiple of the hydrogen nucleus charge? What determines their different masses?

Subatomic physics presents scientists with a whole range of microscopic objects and the common method of investigation is to hit them hard and see what happens. These days it is usually done with charged particles such as electrons, which can be accelerated to enormous energies and speeds close to light in giant electrical machines. In the early days, this technology was not available, and radioactive emissions were the best laboratory source of high-speed projectiles. Later, cosmic rays – extremely energetic particles that strike the Earth from space – also played an important part. The success of bombardment at ever greater energies has been spectacular, and has advanced our knowledge of the microscopic structure of matter enormously. Accelerator machines now in use can impel electric particles on to a target with a force some 10 000 times greater than Rutherford's alpha rays. The resulting impacts burst open a whole new world of subatomic activity which is keeping an army of physicists occupied in trying to comprehend it. Using violent collisions, we can unlock new forces of nature buried in the deepest recesses of matter and expose to view for a fleeting moment a bewildering collection of microscopic inhabitants.

3.1 *A new force*

More than 100 different types of atom (i.e. elements) are known to modern physicists. Recall from Chapter 2 that the chemistry of an atom is controlled by the nuclear charge, which is found to range in whole units from one (hydrogen) up to 106. How can we understand the existence of all these atoms? Why 106?

As so often in science, much progress can be made merely by systematic classification. The vital statistics of nuclear physics are weight (mass), electric charge, magnetic field and spin. In the early days, only the first two quantities were measured in a meaningful way, but already they gave a great deal of information about the nature of nuclei. Fig. 3.1 is a chart of weight versus charge for all the known elements. The first feature to notice is that the heavy nuclei are also the ones with the most charge. Secondly, the line of dots starts out at a gradient of two and curves steadily upwards: the weight starts out being about twice the

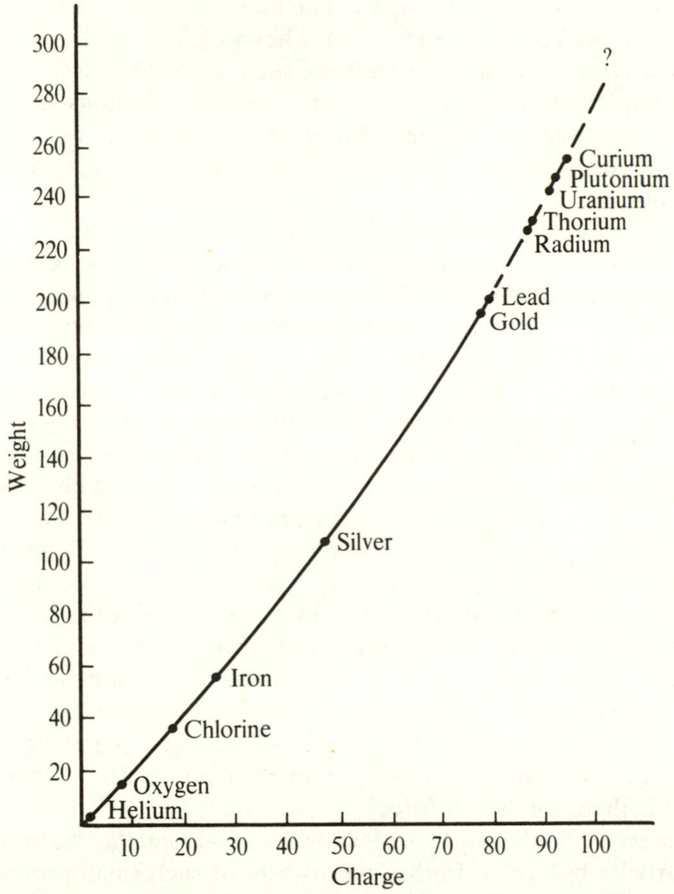

Fig. 3.1. Charge–weight chart for the known nuclei, in units of the charge and weight of the hydrogen nucleus. The broken line represents the region of naturally-occurring radioactive elements. Uranium is the heaviest nucleus found in abundance on Earth.

charge, but slowly becomes somewhat greater.

The fact that the nuclear charge is always a whole number of hydrogen nucleus charges strongly suggests that all the nuclei above hydrogen are really *composite* bodies, containing successively greater numbers of hydrogen nuclei bound together. We now know this to be correct. In 1919 Rutherford and his colleagues were able to displace hydrogen nuclei from the nuclei of heavier atoms by bombarding them with alpha particles. The emerging hydrogen nuclei were released with considerable energy, indicating that the atomic nucleus is a potential power source.

Hydrogen nuclei, being a fundamental building block of all nuclei, can be regarded as elementary particles of matter, as are electrons. No one has succeeded in breaking one apart. The name given to this basic entity is the *proton*. All protons are identical. They weigh about 1836 times as much as electrons and carry, as we have seen, a positive electric charge equal in magnitude to the charge on the electron. Protons share with electrons the property of being *fermions*: they possess an intrinsic spin of one half, and hence obey the Pauli exclusion principle. They also have a magnetic field.

Accepting the existence of protons in nuclei raises two problems. First, protons are all positively charged, so that they repel each other with electric forces. The electric repulsion is an inverse square force (see Section 1.3) so it becomes intensely strong in the crowded proximity of the tiny nucleus, which is only about 10^{-15} m across. A proton in a uranium nucleus, if free to move, would be instantly expelled at the fantastic acceleration of 10^{28} g. How do the protons withstand the disruptive action of these electric forces? Secondly, if nuclei are made *only* of protons, then the line in Fig. 3.1 should lie at 30°; a nucleus with, say, eight protons (oxygen) should be eight times as heavy as hydrogen. Instead it is twice this, with about sixteen proton masses. Higher up the chart the discrepancy is worse, eventually reaching a factor of about 2.7.

Turning to the first problem, one obvious solution is that electricity does not supply the only force that acts between protons. Another, more powerful force must also be operating inside the nucleus to glue the protons together against their electric repulsion. Gravity is far too weak for this (10^{-38} times too weak in fact) so the only possibility is the existence of a *new force of nature* – a nuclear interaction. What other evidence is there for such a force?

Let us recall Rutherford's experiments involving the scattering of alpha particles by nuclei. In this microworld of such small proportions we need to use new units of distance with which to measure. Physicists call 10^{-15} m (about the size of the nucleus) one fermi, after the famous Italian nuclear physicist Enrico Fermi. The fermi is usually abbreviated as fm. Rutherford's scattering of alpha particles from nuclei showed that only electrical forces were operating down to within 10 fm of the nuclear surface. Later experiments made use of protons, accelerated to a higher energy so that they would approach even closer to the surface of the nucleus, and a study of their scattering characteristics showed that there is indeed a powerful new force that starts to disturb the protons' trajectories when they approach to within 2 or 3 fm of the nucleus. Clearly, the nuclear gluing force is strong, but of very short range – quite different from the more familiar electromagnetic and gravitational forces, which can operate over macroscopic distances. The exceedingly short range of the nuclear force explains why we do not observe it in daily life.

The discovery of a new force of nature is of the greatest signficance, and it increased the number of known types of interaction between matter from two to three. Earlier hopes of unifying electromagnetism and gravity into a fundamental theory of all interactions soon dissolved when it was found that this third force existed and, moreover, appeared to be qualitatively very different from the other two.

Let us now turn to the second question – why aren't the masses of nuclei proportional to their electric charge? The excess masses can obviously be explained if we suppose that in addition to the protons, the nucleus contains other matter, in the form of electrically *neutral* particles. This conjecture, assumed by Rutherford as early as 1920, was verified in 1932 by the British physicist James Chadwick (1891–1974), who succeeded in displacing neutral particles from the nuclei of beryllium by bombarding them with alpha rays. He received the 1935 Nobel physics prize for this discovery. The particles were called *neutrons*, and are found to have a mass almost exactly the same as that of the proton. A careful measurement reveals that it is just a little greater – a fact of some significance as it turns out. Although neutrons exert no electric force, the fact that they are confined to the nucleus implies that they must be subject to the nuclear force which binds the protons and neutrons together.

The discovery of a nuclear force and a neutral particle immediately provides an explanation for the more obvious characteristics of nuclei. Consider, for example, the oxygen nucleus. Fig. 3.1 reveals that it contains eight units of charge and about 16 units of mass. We therefore conclude that the nucleus consists of eight protons and eight neutrons glued together (see Fig. 3.2). However, a close examination of terrestrial oxygen reveals that not all oxygen nuclei are 16 times as heavy as hydrogen. About 0.2% is 18 times as heavy, and a tiny 0.04% is 17 times as heavy. These atoms are all *chemically* oxygen, so all must have eight protons in their nuclei, but they differ in their weight. Such sibling

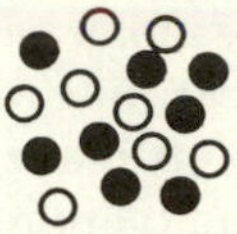

Fig. 3.2. The composite nucleus. Atomic nuclei, such as this one of oxygen, consist of two types of particle, neutrons (unshaded) and protons (shaded), bound together by strong nuclear forces. Each particle contributes roughly equally to the total mass, but only the protons are electrically charged. In most oxygen nuclei there are sixteen particles, so these nuclei are about sixteen times as heavy as a hydrogen nucleus (one proton). However, a small proportion has two extra neutrons, and a still smaller proportion one extra neutron. These are called *isotopes*. Oxygen isotopes with six, seven and eleven neutrons can also exist, but they are unstable, and decay radioactively within a few minutes.

atoms are called *isotopes*; all nuclei are now known to have at least a small number of isotopes. Their existence is easily explained: they differ in the number of neutrons that they contain. Terrestrial oxygen nuclei can possess either eight, nine or ten neutrons.

Much remains to be explained. What determines the number of neutrons allowed in nuclei? Why doesn't oxygen occur with 23, or 5000 neutrons, or none at all? What role do they play and why are proportionately more of them present in the heavier nuclei, for example uranium? Why don't we encounter free neutrons outside nuclei?

All these questions, which can be readily answered, have to do with nuclear *stability*. The existence of every nucleus is a competition between powerful opposing forces: electric repulsion and nuclear attraction. If there is an overbalance, the nucleus is threatened with disaster, and can disintegrate violently. Many examples are known of spontaneous disintegration. Elements such as uranium and radium are found to eject energetic charged particles from their nuclei in an attempt to reorganize the delicate nuclear balance of forces. This phenomenon is called *radioactivity*, and is the source of, among other things, alpha rays. The reason why some nuclei are radioactive will be discussed in the next section.

Most naturally-occurring radioactive nuclei are clustered towards the top end of the curve in Fig. 3.1, and have been marked with a broken line. It suggests that the absence in the Earth of heavier elements than uranium is due to the fact that if they existed they would be radioactive with such short lifetimes that they would have disintegrated long ago. This has indeed been confirmed, for although they are not found in the Earth, physicists can *make* these heavier 'transuranium' nuclei artificially in the laboratory, by bombardment of uranium and other heavy elements with subatomic particles. At the time of writing 14 new heavy elements have been made this way, the heaviest of which contains 106 protons and 157 neutrons. It has a lifetime of 1 s. As expected, none of the transuranium elements survive as long as the age of the Earth (4.5 billion years), though some of them, such as plutonium, live long enough (76 million years) to be useful as nuclear fuels. (There is some evidence that a minute amount of natural plutonium may be present on Earth.) Apart from plutonium the greatest longevity, about 16 million years, is achieved by an isotope of the element curium which contains 96 protons and 151 neutrons. The heaviest naturally-found element, uranium, has a lifetime about the same as the age of the Earth, which is about 300 times longer than the lifetime of curium. Thus, the absence of natural transuranium elements is explained.

A study of radioactive emissions reveals that the expelled alpha particles are also composite bodies, consisting of two protons and two neutrons united together. This combination is, in fact, the same as a nucleus of a helium atom. Whenever a uranium nucleus decays by emitting an alpha particle, it follows that it loses two protons and two neutrons. The loss of the two protons reduces the nuclear charge by two, and therefore converts the uranium into a completely different chemical element (thorium in fact). The idea of one element transmuting into another obsessed the early alchemists (whose interest lay chiefly in producing gold) and was long held in derision by scientists. The alchemists were right after all, but their procedure – of trying to alter the elements chemically – was doomed from the start, because chemistry involves rearrangements of just the outermost atomic electrons. Only the much more powerful nuclear rearrangement can achieve transmutation.

Many of the radioactive elements lighter than uranium (e.g. radium) have much shorter lifetimes, but their existence on the Earth can be explained. They are the *decay products* of heavier nuclei with longer lifetimes, such as uranium, and have been produced by their recent disintegration. The elements below bismuth (83 protons) can all exist in stable form. However, radioactive isotopes of all the elements can be made artificially, the first having been produced by Irène Joliot-Curie (French, 1897–1956) and Frédéric Joliot (French, 1900–58) in 1934, for which they received the 1935 Nobel chemistry prize. Today about 1500 are known. All have short lifetimes, which accounts for their almost complete absence in nature. One exception is the isotope of carbon with eight (rather than the usual six) neutrons. This is found in the atmosphere because it is continually being produced from atoms in the air by cosmic ray bombardment. With a lifetime of 5730 years, it is useful for the radioactive dating of once-living materials.

The phenomenon of radioactivity is the reason why only about 100 elements occur in nature, but it is not an *explanation*. We need to know why the heavy nuclei, and the neighbouring isotopes of the stable light ones, are unstable. To understand nuclear stability, and so explain the systematic features of all these statistics, we must examine the nuclear balance and weigh up the competing forces. Even a crude model explains the general features very well.

Imagine the nucleus as a ball of protons and neutrons. Each nuclear particle – or nucleon as it is called – attracts its neighbours with the strong, short-ranged nuclear force. If the nucleons become too dispersed, the protons would start to move out of the range of the force, and would explode away under their electric repulsion. Only when they are within a distance of about 1 fm of each other are the nucleons safely gripped by the nuclear glue. Now if each nucleon attracts its neighbour, and that in turn attracts its neighbour, and so on, it seems that all the nucleons must be irresistibly drawn together into a tight ball less than 1 fm across. There is much evidence that this does not happen.

Suppose the nucleus was collapsed, with all the nucleons moving within the force range of *every* other one, then every particle would attract every other. If now another particle were added to the nucleus, more attractive bonds would appear and tighter gluing would result. Thus the most stable nuclei would be the heavy ones, because these have the most nuclear bonds to cohere them. But this is in direct conflict with the facts about natural radioactivity, which show that the heavy nuclei are actually the *least* stable.

Direct information about the internal arrangement of nuclei can be obtained from experiments in which high-speed electrons are shot right *through* nuclei. Electrons are not affected by the nuclear force, so they do not feel the neutrons, but they can probe the organization of the protons by responding to the electric fields associated with them. It is found that the *density* of protons is more or less the same in all nuclei – they are not packed substantially more tightly in the heavy than the light ones. The greater the number of nucleons, the larger is the nucleus. Very roughly, the volume of a nucleus is proportional to its total particle content, showing that the nucleons are not all collapsed together in a tight crowd, but are spaced out about 1 fm apart. From this observation, it follows that the nuclear material must have a certain amount of *rigidity*: in short, it is extremely sticky, but it can't be squashed too much.

The reason for the rigidity of nuclear matter will be discussed later, in Section 4.2, but for now we simply note its improtance for the stability of the nucleus. Because of the relatively low population density not all the nucleons attract each other; only the *nearest neighbours* are within range. Clearly, a nucleon in a heavy nucleus such as uranium is glued no more tightly than one in a light nucleus such as oxygen, since in each case it only sticks to the nearby particles.

How can this important conclusion be checked experimentally? It is possible to do this by very careful measurements of the relative weights of nuclei. First, a technical detail must be considered. The weight of a nucleus is in the region of 10^{-25} kg, which is an inconveniently small quantity. A more useful unit which is widely used is the electron volt, written eV, corresponding to the energy acquired by a single electron that is accelerated in an electric field through a potential difference of one volt. The mass equivalent of this energy (via Einstein's relation $E = mc^2$) is about 10^{-36} kg. In these units the electron has a (rest) mass of 0.511 MeV. The proton and neutron masses are 938.28 MeV and 939.57 MeV, respectively.

Consider now the oxygen nucleus with eight protons and eight neutrons. It might be expected that it should weigh (in MeV)

$$8 \times 938.28 + 8 \times 939.57 = 15\,022.8.$$

In fact it weighs only 14 895 MeV, or about 128 MeV less. This deficit, equivalent to the mass of about 250 electrons, is getting on for 1% of the total. What has happened to the missing mass?

The answer is provided by the special theory of relativity, and the famous equation $E = mc^2$ expressing the equivalence of mass and energy. Suppose we tried to pull a nucleon out of the oxygen nucleus. To overcome the grip of the nuclear force we would have to work hard and supply a great deal of energy, about 8 MeV in fact, to break the bonds. Conversely, if the nucleon were absorbed back in the nucleus, 8 MeV of energy would be released. This applies to all the nucleons of oxygen. If we were to pull the whole nucleus to bits, a total of about 128 MeV would be needed. This energy is referred to as the binding energy of the oxygen nucleus. Consequently, the oxygen nucleus would emit this quantity of mass-energy if it were assembled again.

Not only does $E = mc^2$ explain why the assembled nucleus is lighter than the sum of the separate masses of the individual particles it contains, it also enables us to calculate the binding energy by careful measurement of the nuclear mass. If we turn to a very heavy nucleus, such as uranium with 238 nucleons, we find a mass deficit of about 1800 MeV, which is once again getting on for 1% of the total mass. That is, it only requires about the same 8 MeV of energy to remove a nucleon from uranium as it does from oxygen. This confirms that the binding in heavy nuclei is no stronger than in light ones.

In very light nuclei, such as helium, the number of nucleons is so low that they do not use up even the limited number of bonds available to them (not enough neighbours around), so these nuclei are rather less strongly bound than the medium weight ones (see Fig. 3.3). This fact is

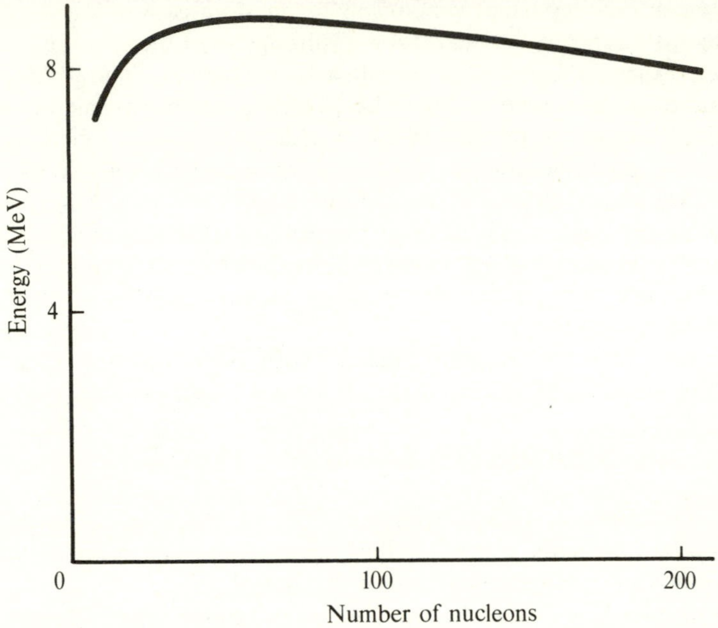

Fig. 3.3. Binding energy of nuclei. This graph shows the energy needed to pull one nucleon out of each of the known nuclei. It is nearly constant over a wide range, indicating that only nearest-neighbour binding forces operate. The very light nuclei are weakly bound because of a paucity of neighbours. The very heavy nuclei are less strongly bound than the medium ones because of their additional electric repulsion. Thus both fusion of light nuclei and fission of heavy nuclei produce more tightly bound progency, and so release energy.

of great significance for the evolution of stars, the energy output of which depends on nuclear reactions involving light nuclei.

It is worth comparing the binding energy of a nucleon in a nucleus with that of an electron in a hydrogen atom. The energy required to release the electron from the grip of the proton is about 0.0027% of its rest mass. These figures are tiny against the 1% for the nucleon, and indicate that the energy involved in nuclear rearrangements is enormously greater than that of chemical energy, which is why a nuclear bomb using 1 kg of uranium is so much more destructive than 1 kg of TNT.

The fact that the nuclear cohesion does not improve with nuclear size is of vital importance to the nuclear balance because the electric force, which is pushing out against the binding, is *not* restricted to nearest-neighbour interaction. Electric forces, unlike nuclear forces, are long-ranged, and operate right across the nucleus. The cumulative effect of all the electric repulsion goes on getting bigger as we consider heavier nuclei with more and more protons. But the fight is not completely one-sided. The nuclear force can call on the neutrons for help, because they are strongly attractive from their nuclear force, but do not feel any electric disruption at all because they carry no electric charge. It is now possible to explain why the line in Fig. 3.1 curves upwards: stability in heavy nuclei favours neutron richness to dilute the protons and contribute extra glue to overcome the mounting electric pressure.

So what goes wrong in the very heavy nuclei?

The explanation of these facts is the first indication that quantum mechanics plays a vital role inside the nucleus as well as outside.

3.2 *Radioactivity*

The behaviour of heavy nuclei containing perhaps more than 200 interacting particles is very complicated, and it is necessary to resort to a number of idealized models. In some ways the heavy nuclei behave like a box of gas or a drop of liquid, in which the individual particles do not play a conspicuous role. In other ways they display some of the properties seen in atomic physics with an ordered energy shell structure. Or, in certain situations, internal particle-like behaviour may be very important, with certain groupings of protons and neutrons in evidence. It all depends which processes or properties are under study.

Regarding a very heavy nucleus such as that of uranium as like a drop of liquid gives insight into why nuclei with substantially more than 100 protons cannot exist. It is a familiar fact that small drops of water tend to contract into spherical balls. The explanation for this is easy to understand (see Fig. 3.4). The molecules of water are attracted to each

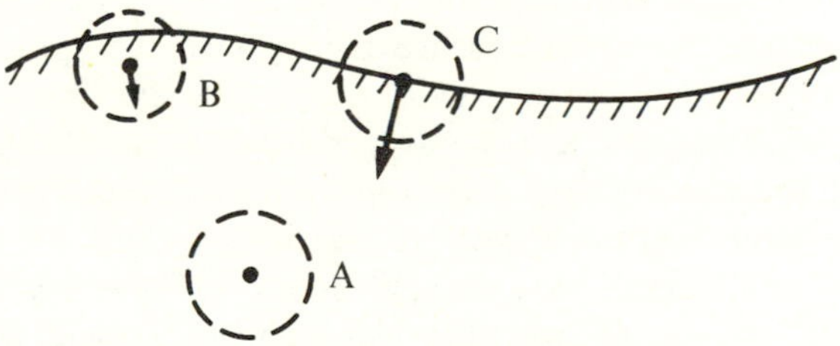

Fig. 3.4. Surface tension. Molecules (black dots) feel the cohesive forces due to their nearby neighbours inside the region denoted by the broken circles. At location A these all balance out, and there is no net force. At B there is a net inward force due to the unequal distribution of liquid round the molecule. This force rises to a maximum at the surface C, where all the cohesive attraction is directed inwards. Consequently, if we were to transport a molecule from position A to position C we would have to work against this force and expend energy. This means that whenever the surface area of the liquid is increased – by a change of shape – energy must be supplied to populate the extra surface with molecules from the interior. Therefore the liquid tries to adopt a shape which minimizes the surface area.

other by short-ranged, weak electromagnetic forces. They are short-ranged because they are of the dipole type discussed on p. 11 in connection with forces between magnets. A molecule near the middle of a drop is held by the cohesion of all the other nearby molecules which surround it on all sides. In contrast, a molecule very near the surface of the drop only feels a pull towards the direction of the interior, simply because being near the surface most of its neighbours are on one side of it. The result of this lopsided attraction is to pull the surface layers inwards somewhat, creating a sort of taut skin effect, known as surface tension, which tries to squeeze the droplet.

Simple energetics tells the rest of the story. If we wish to remove a molecule from deep in the interior of the drop, we would have to supply enough energy to overcome the restraining forces of all the surrounding molecules, but to remove a molecule from the surface only requires roughly half of this, because a surface molecule is only restrained by the neighbours on one side of it. If the droplet is spherical, it requires the expenditure of energy to deform it, because extra surface area – and hence less cohesive attraction – is thereby produced. The surface tension tends to pull the droplet into a spherical shape to minimize its energy. A large blob of water on a level surface will deform into a pancake shape because it is disrupted by gravity. The loss in energy by the blob in forming a pancake with a large surface area is compensated by the energy gained by lowering its centre of gravity.

In the nucleus of an atom there is also a surface tension which tends to make the nucleus spherical in shape. This time the cohesive force is the strong nuclear one. There is in addition a disruptive force, though of internal rather than external origin: the electric repulsion. If the nucleus deforms slightly, the protons in the bulges move a little farther from the protons near the centre, which are trying hard to repel them electrically – so there is an energy gain from this. In compensation, the alteration in shape of the nucleus produces some additional surface area, which reduces the total cohesion somewhat because some internal particles become additional surface particles (which are less strongly bound). In small nuclei, the surface energy wins every time and the deformation is squeezed away. But as we consider nuclei with more and more protons in them, so the electric repulsion grows in strength relative to the surface tension. The reason for this is given in the box on p. 95.

An elementary calculation, based on this simple analysis, taking into account the relative strengths of the electric and nuclear forces, shows that a nucleus stands to gain an energy bonus by deforming, so long as it contains at least about 140 protons. A larger nucleus than this would therefore become unstable and break into two or more smaller pieces, which would have a lower total energy than the larger single nucleus.

The nuclear balance: surface attraction energy versus electric repulsion energy
As the nuclear force is short-ranged, the nuclear particles do not tend to crowd near the centre much, but remain more or less uniform in density. Thus the volume of the nucleus is proportional to the total number of particles, say N. The radius is therefore proportional to $N^{1/3}$ and the surface area to the square of this, so the surface energy is proportional to $N^{2/3}$. In contrast, the electric repulsion energy is proportional to (charge)^2/(charge separation). The charge is determined by the total number of protons, which for heavy nuclei is always about 40% of N. The average separation of protons is about a nuclear radius, which is proportional to $N^{1/3}$. So

$$\frac{\text{electrical energy}}{\text{surface energy}} \propto \frac{(0.4N)^2/N^{1/3}}{N^{2/3}} \propto N.$$

Thus, for heavy nuclei, with large N, the electrical energy dominates and the nucleus becomes unstable against fission.

This phenomenon of spontaneous nuclear fission actually occurs in nuclei somewhat lighter than 140 protons if they are disturbed in some way, such as by neutron bombardment. This is the principle used in nuclear power stations, where the energy liberated by uranium fission is used to make electricity.

In addition to instability against large-scale fission, heavy nuclei are even more unstable against the emission of a small conglomerate of nuclear particles, namely a helium nucleus. These emissions are the alpha rays discussed in the previous chapter. It is here that another aspect of nuclear behaviour shows its face, in which the particle nature of its interior is important. In some ways, the internal structure of the nucleus behaves as though the particles it contains are not moving about as individuals, but grouped together into alpha particles. This grouping turns out to be a particularly favoured one.

The reason that a small piece of the heavy nucleus is more likely to break off than a large piece is now well understood, and an analysis of the process of alpha radioactivity nicely illustrates the importance of quantum phenomena in the internal nature of the nucleus. When alpha rays were first being studied by Rutherford, it was discovered that uranium nuclei emitted these particles with a definite energy. However, when uranium was bombarded from another source with alpha rays which were more than twice as energetic, none of these particles seemed to be absorbed back into the uranium nuclei. On the reasoning that if an energetic alpha particle cannot get into a uranium nucleus, a less energetic one should not be able to get out, a mystery surrounded the mechanism of alpha escape.

This mystery is graphically illustrated by considering the experience of an alpha particle as it is brought up close to the nucleus. The positive electric charge of the nucleus fiercely repels the approaching alpha particle because of the two protons it contains, and the force increases as

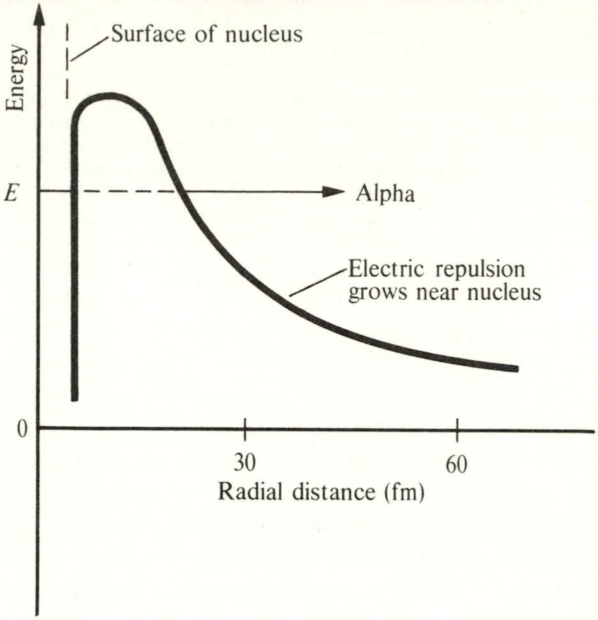

Fig. 3.5. Potential barrier by an alpha particle around a uranium nucleus. The central well is due to the average nuclear attraction of all the nucleons and the hill is due to the electric repulsion of the protons. Alpha particles with energy E trapped inside the nuclear well may still escape to become alpha rays, by quantum mechanically tunnelling through the barrier.

the nucleus is approached. Suddenly, at the nuclear surface the short-ranged nuclear attraction is felt, and as it is much stronger than the electric repulsion, it overwhelms the latter and draws the alpha particle into the nuclear material in spite of it. The pattern of force surrounding the nucleus thus has the form of a so-called 'potential hill' shown in Fig. 3.5. This graph plots the energy needed for the alpha particle to reach a given radius as measured from the centre of the nucleus, so it is rather like the cross-section through a hill, where energy is needed to climb the slope (against gravity rather than electric forces in that case), surmount the hump, and drop into the valley below. Whether you are climbing into, or out of the valley you still need enough energy to get over the peak.

The mystery with Rutherford's alpha particles is that they have nowhere near enough energy to climb out of the well. How, then, do alpha particles get out? According to classical ideas about energy they would be forever trapped in the well. However, quantum mechanics allows energy conservation to be violated for very short durations according to the energy–time uncertainty principle (2.8). For about 10^{-24} s the alpha particle can acquire enough energy to rise to the level of the hill top, and this is just enough time for it to travel sufficiently far from the nuclear surface to escape the powerful grip of the nuclear force. So by the time it has to pay back the borrowed energy, it finds itself on the outside slope of the hill, free of the nuclear force. It is then impelled clean away by the electric repulsion. To all intents and purposes, the particle behaves as though it has tunnelled *through* the potential hill.

Using a full-blooded calculation, it is possible to compute in a simplified model the probability that a nucleus will emit an alpha particle by quantum tunnelling. The result is extremely sensitive to the energy of the emitted particles. For the nucleus of thorium containing 232 nucleons it takes an average of 14 billion years for an alpha particle of 4 MeV energy to escape. On the other hand, a polonium nucleus with 213 nucleons which emits alpha particles only about twice as energetic as this, decays in a mere 4 μs!

The quantum mechanical nature of alpha radioactivity explains why radioactive substances have a definite lifetime. According to quantum ideas, each alpha particle penetration is a purely random event with a well-determined *probability*. In the case of uranium, each nucleus has a 50-50 chance of decaying in about 4.5 billion years, so on average, about half the nuclei in a lump of uranium will have disintegrated after that duration. Of the surviving 50%, one half of that will decay after another 4.5 billion years, and so on. Each 4.5 billion years the uranium content divides by half, so this time span is known as the *half-life*. Whilst any individual nucleus may take a much shorter or longer time to disintegrate, the half-life is a good measure of a typical decay time. It turns out that this pattern of decay with a definite probability and half-life applies to all subatomic transmutations – an indication of the rule of quantum mechanics over all microphysical activity.

As indicated in Fig. 3.1, alpha radioactivity occurs among the very heavy nuclei above lead (82 protons). However, artificial alpha-active nuclei can be made which are lighter than this, though most of them are very unstable. These are all isotopes of otherwise stable elements which have been deliberately made deficient in neutrons, and alpha decay helps redress the balance. Even so, with the exception of a few very light nuclei (for which alpha disintegration really amounts to spontaneous fission into two nearly symmetric pieces) and an isotope of chlorine, none is known below an isotope of the element neodymium (60 protons). Most other artificial radioactive isotopes decay by a different process altogether, called *beta decay*, in which an energetic electron (called a beta particle) is ejected from the nucleus. It will be discussed in detail shortly. Natural beta radioactivity is also found competing with alpha decay among the heavy nuclei. A study of beta decay has consequences of great importance for fundamental physics, revealing the existence of yet another new force of nature. In addition, it provides important insight into the quantum mechanical organization of the nuclear interior.

It is easy to understand why nuclear stability is threatened by neutron deficiency: the neutrons assist in sticking the protons into the electrically repulsive nucleus. Less obvious is the reason why a neutron excess is also undesirable. An analysis of natural, stable nuclei shows that their isotopes all lie in a very narrow range of neutron content; usually only a variance of one or two either way. Artificial nuclei lying outside this narrow band are unstable, and disintegrate, usually via beta decay. Why do neutron-rich nuclei become unstable? Taking it to the extreme, why aren't there any nuclei built entirely out of neutrons, and why do we not find individual neutrons in nature?

These questions can be answered by observing the fate of neutrons that are displaced from nuclei by bombardment. The result is a phenomenon that requires a completely new theory of microscopic matter to explain. After a half-life of just over 15 minutes, the neutron suddenly disintegrates, ejecting an electron and a proton. What can be made of this extraordinary phenomenon?

The first idea that comes to mind is that a neutron is really a tightly bound union of a proton and electron which when freed from the confines of the nucleus somehow becomes unstable and explodes apart. However, there are good reasons why this explanation is unacceptable. Quantum theory shows that the lowest electrically bound state of the proton–electron system is the ground state of the hydrogen atom. To confine an electron as part of a neutron into the nucleus of an atom of only a few fm in size would not only require a new, stronger, type of force for which the scattering of electrons from protons or nuclei gives no evidence, but would also, through the uncertainty principle, impose on the electron an energy about 40 times its own rest mass. It would move with a speed of 99.97% of the speed of light. It is hard to imagine such entities smashing around inside the nucleus without leaving some sort of trace.

There is another reason why the neutron cannot be a composite of electron and proton. It is found that the neutron shares with the latter two particles the property of possessing intrinsic spin one half (see Section 2.7). The only ways that the electron and proton can combine their half units of spin is to give either one unit or zero, neither of which matches that of the neutron.

The picture which modern physics paints of the decay of the neutron is much more profound than a simple fission. Rather than envisaging the particle as simply coming apart, we now think of it as *disappearing* altogether from the universe – ceasing to exist at a particular point of space–time – and an electron and proton being created where none existed before. A proper understanding of the creation and destruction of particles of matter requires a deeper insight into the role of relativity and field theory in quantum affairs, and this will be dealt with in the coming sections. For now we shall briefly see how the existence of neutron decay explains beta decay and the balance of forces inside the nucleus.

It is possible to construct a mathematical model of the nucleus in accordance with quantum laws along the lines of the quantum model for the atom, discussed in Chapter 2. The difference is that instead of orbiting around a more or less fixed centre of attraction, as is the case with electrons, the neutrons and protons are all intermingled in a blob, with nuclear and electric forces all around them. In addition, the nuclear force itself is not a simple inverse square law, as is the case for the atomic electrons, but a very complicated type of force which is still not completely understood. As well as depending on the location of the nucleons, it also depends on their relative spin orientations. These features make the mathematical description of the nucleus much more complicated than that of the atom, and only limited progress has been made. Nevertheless, some basic features are well understood. In particular, *energy levels*, similar to those of the atomic electrons, are predicted theoretically and also observed experimentally.

The existence of nuclear energy levels is of crucial importance for nuclear structure and stability, for the following reason. As mentioned briefly on p. 98 the proton and the neutron share with the electron an all-important property: intrinsic spin one half. This means that protons and neutrons are *fermions* (see p. 80) and so are subject to the Pauli exclusion principle. Consequently, just as electrons cannot enter already occupied levels and must stack up above one another in the higher levels, so too does the nucleus contain proton and neutron stacks. As the neutrons are distinguishable from the protons, the stacking rules operate quite separately (a proton and a neutron could co-habit the same level if they happen to coincide). Each group of nuclear particles stack up in their separate levels from the bottom upwards.

Although similar in structure, the proton energy levels start off higher than their neutron counterparts because of the additional energy which must be supplied to the protons to confine them in the nucleus against their electric repulsion. This energy gap grows with the size of the nuclei, for reasons explained on p. 92. In stable nuclei, it is found that the *tops* of the two stacks are about level, so that more neutrons than protons are present in the heavy nuclei to fill up the gap.

One fact which needs an explanation is why the neutrons in a nucleus do not decay when isolated neutrons have a half-life of only a few minutes. We need look no further than the Pauli principle to answer this question. To decay, a neutron would have to change into a proton and an electron. The electron, which is not subject to the strong nuclear force, would be rapidly ejected from the nucleus, but the proton would remain behind and would have to be located in one of the proton energy levels. As all the lower levels are full, it can only be accommodated at the top of the stack, but in general the disintegration energy will be insufficient to enable the proton to reach this elevated location. The neutron decay will then be unable to proceed simply for lack of energy. The neutron would 'like' to decay, but is forbidden to do so by the law of energy conservation.

In nuclei that have rather too many neutrons, the neutron stack is significantly higher than the proton stack. It is then energetically favourable for a neutron at the top of the pile to decay and place the resulting proton at the top of its pile. In this situation the nucleus is unstable, and will eject an electron, thereby converting a neutron into a proton. This rearrangement brings the relationship betwen the neutron and proton piles back into balance. When it happens, the ejected electron is known as a beta ray or beta particle, and the phenomenon is called beta radioactivity (see Fig. 3.6). As with alpha radioactivity, beta emission alters the nuclear force balance by changing the electric charge. By acquiring an additional proton, the nucleus moves *upwards* in the list of chemical elements. Note that the total electric charge remains the same. The ejected electron carries an equal and opposite (negative) charge which exactly balances the positive charge on the newly created proton, in accordance with the law of electric charge conservation.

The exclusion of neutrons (and protons) from the same quantum energy levels has the effect of slightly altering the net nuclear attraction between like particles as compared to the neutron–proton attraction. Very roughly, a neutron trying to enter a filled lower energy state would be repelled by other neutrons with the same spin direction. It is a fascinating thought that this neutron repulsive effect is also responsible for preventing highly compact stars, known an neutron stars, from collapsing under their own weight.

Because of the altered nuclear attraction, energy saving favours *symmetric* nuclei with equal numbers of protons and neutrons, for these provide the greatest number of attractive neutron–proton bonds for the least amount of neutron–neutron and proton–proton exclusion repulsion. In light nuclei, where the electric forces are unimportant compared

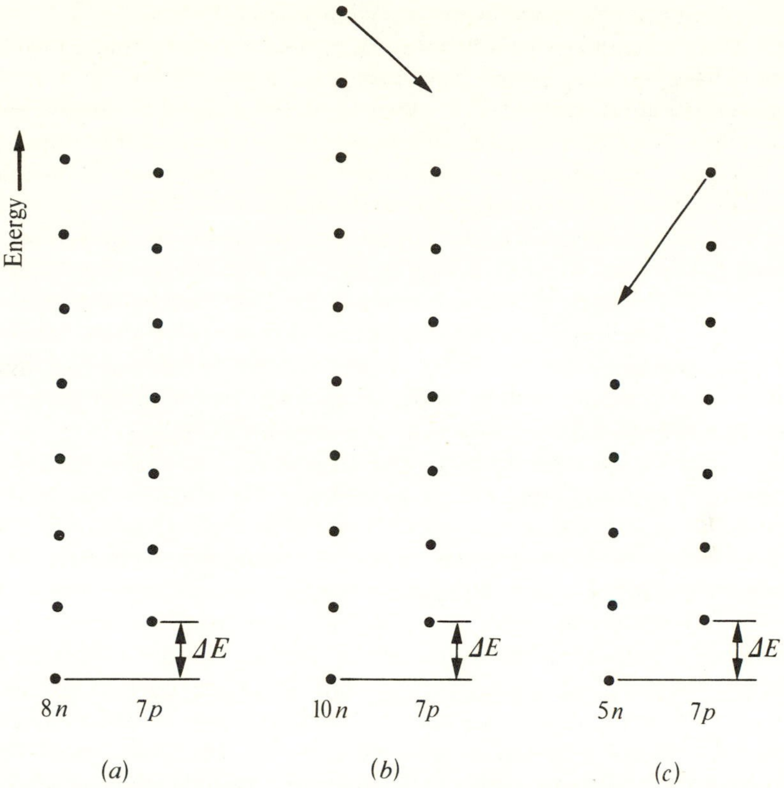

Fig. 3.6. Beta radioactivity. (*a*) Stable nucleus (nitrogen isotope) in which the tops of the neutron (*n*) and proton (*p*) stacks are about level. The gap ΔE at the bottom of the proton stack is due to their electric energy. This gap is why most stable nuclei have rather more neutrons than protons. (*b*) Unstable isotope of nitrogen containing too many neutrons. There is an energy bonus from converting one into a proton via beta decay (emission of an electron: half-life 4.2 s) to form a stable isotope of oxygen. (*c*) Nitrogen isotope with proton excess. Energy considerations favour proton conversion into a neutron via beta decay (emission of a positron; half-life 0.011 s) to form an isotope of carbon.

with these nuclear effects, we do indeed find equal numbers of protons and neutrons. For example, the most abundant isotope of oxygen has eight of each. As far as the heavier nuclei are concerned it is clear that as the proton number increases (and electric repulsion mounts) a balance has to be struck between the conflicting requirements, on the one hand, to add more and more neutrons to dilute the protons and contain the electric disruption, while on the other hand to attempt to equalize the numbers of protons and neutrons. It is the point of balance of these opposing tendencies which determines the narrow range of neutron numbers found in stable nuclei. Too many neutrons, and the symmetry energy favours a neutron conversion to a proton via beta decay. Too few neutrons, and symmetry now dictates the *reverse* process – that a proton should change into a neutron. Can this also happen?

All fundamental physical processes can be reversed. If a neutron can decay into a proton and an electron, then we must be able to reconstitute a neutron again from its decay products. The reason that the free neutron can decay depends on the fact that it has a greater mass than the proton and electron combined. Its decay is therefore a 'downhill' process, favoured on energy grounds according to the relation $E = mc^2$. The reverse process is 'uphill' and will not happen unless energy is fed in from outside. This is fortunate otherwise there would be a natural tendency for hydrogen atoms to convert themselves into neutrons. Then there would be no stable stars and no life in the universe (stars like the sun burn hydrogen in nuclear furnaces).

It is, however, possible to make free neutrons by colliding electrons and protons with sufficient energy to make up the mass deficit. In the nucleus, high-energy electrons are not available, but it turns out that they are not needed: the *nuclear* energy can make up the deficit. The energy gain when a neutron-deficient nucleus changes a proton to a neutron more than offsets the mass discrepancy between the two. There is a ready supply of available electrons orbiting around the nucleus although they are, of course, very remote from the nuclear surface. However, in quantum systems one must always allow for the inherent uncertainty in the position of all electrons, so there is a small but significant probability that one of the innermost orbital electrons will be found *inside* the nucleus for a brief moment (see p. 71). If that happens then the electron can be captured by one of the nuclear protons, and a neutron thereby produced. This process is particularly favoured in heavy atoms, where the nuclei are large and their high electric charges pull in the inner electron orbits. Later we shall see that there exists another mode of transmutation (positron emission) which enables a proton to change into a neutron.

These considerations will now be illustrated with an example. Lead is a heavy element whose nuclei contain 82 protons and anything from 112 to 132 neutrons. It has four stable isotopes: those containing 122, 124, 125 and 126 neutrons. All the neutron-deficient isotopes (which can be made artifically) can decay by capture of an orbital electron, to form a nucleus of thallium with only 81 protons. Thus, lead with 116 neutrons becomes thallium with 117 neutrons, after a half-life of 2.4 hours. With 120 neutrons, stability is less threatened, and the half-life is as long as 300 000 years; for 123 neutrons it reaches 30 million years. The neutron-rich isotopes, on the other hand, decay by electron emission and convert to isotopes of bismuth, which has 83 protons. Most have half-lives of only a few hours. Another example is illustrated schematically in Fig. 3.6.

Although the fact of beta decay explains the distribution of particles in the nucleus, an examination of the details raises a question about its mechanism. The ejection of an electron from the nucleus can be carefully studied, and it is found that the energy of the electron varies over a *continuous* range up to some maximum value, in spite of the fact that the total energy available for beta decay is the same in all nuclei of the same type. If we regard the ejection as rather like a bullet from a gun, simple mechanics requires that the energy of ejection and recoil are always the same. There is a mystery of explaining where the missing energy has gone in all the cases where electrons are ejected with less than the maximum available energy. The answer was suggested in 1933 by Pauli. The missing energy, he reasoned, must be carried away by a *third*, unknown, particle which had up to that time been overlooked. Because of variations in the angle between the two ejected particles, electron and unknown, the electron could come out with a whole range of energies (as observed) with the mystery particle taking up the slack.

The elusive new particle was called the *neutrino* (meaning 'little neutral one') and for some years was not observed directly, due to its extraordinary lack of tangible properties. Careful measurement shows it to carry no electric charge and to be completely unaffected by the strong nuclear force also. Moreover, it is believed, like the photon, to have no rest mass, so is destined always to travel at the speed of light. It is found to possess intrinsic spin one half, so it is a *fermion*. In some ways it is rather like a massless, chargeless electron. So insubstantial are the neutrinos that they can easily pass through the Earth without stopping. Indeed, most would pass through thousands of light years of solid lead!

Measurements show that if the proton were imagined as a perfect absorber of neutrinos, the effective cross-section that it presents would be no larger than a few times 10^{-48} m^2, a billion billion times larger than the area it presents to the neutron. Needless to say, directly detecting neutrinos in the laboratory is difficult, but nuclear reactors produce them in such prolific quantities that at least a few are stopped in ordinary material. By searching for events which are the reverse of beta decay, a direct verification of their existence has been obtained. Neutrinos play a vital role in many astrophysical situations.

The picture which we now have of beta decay is that of a neutron disappearing and a proton, electron and neutrino taking its place. However, as will be recalled from p. 102, there is also the related process in which a proton captures an electron and turns into a neutron. Clearly this inverse process is not the exact reverse of neutron decay, for the latter would require the simultaneous encounter of all three particles: proton, electron and neutrino, which is not only unthinkably improbable but the exceedingly weak interaction between neutrinos and matter would require countless billions of these triple encounters before a neutron is likely to be constituted.

An alternative explanation for the conversion of a proton and an electron to a neutron must be found. The solution lies in the existence of another kind of neutrino, which is *emitted* when the neutron is formed. This new neutrino plays a kind of reverse role to the other one – its ejection is equivalent to the other's absorption. As we shall see in the next section, this mirror relationship is an example of a very fundamental duality of all matter, and is appropriately described by calling one of these other invisible particles an *antineutrino*. For historical reasons, the name antineutrino is given to the particle emitted during neutron decay, the word neutrino applying strictly to the one which is emitted during neutron reconstitution. Often 'neutrino' is used as a general term to encompass both varieties.

The situation is now a little complicated, so it is helpful to illustrate it using letters to denote the particles: n = neutron, p = proton, e^- = electron, ν = neutrino, $\bar{\nu}$ = antineutrino. The decay of the neutron is thus represented symbolically as

$$n \rightarrow p + e^- + \bar{\nu}$$

while the other related process in which a proton changes into a neutron is represented by

$$p + e^- \rightarrow n + \nu.$$

The latter process is clearly different from the reverse of the first, which reads

$$p + e^- + \bar{\nu} \rightarrow n.$$

3.3 *Antimatter*

Until now, we have regarded the forces of nature – electromagnetism, gravitation and the nuclear interaction – in a rather classical way, with the forces manifesting themselves by disturbing the motions of material particles, after the fashion of Newtonian mechanics. Quantum mechanics has played a small role, through the Pauli and the Heisenberg principles, but the essential idea remains of an interaction between separated bodies acting to force them away from freely-moving, straight-line paths. In the next two sections the notion of *interaction* will be extended to a considerably more profound and powerful concept. The necessity for this is forced upon us by the existence of such processes as the decay of the neutron, in which a particle of matter suddenly disappears from the universe, and other particles of a completely different nature suddenly appear. How can such phenomena, in which particles can come and go from the physical world, be encompassed within our theories of physics?

One of the most disturbing aspects about particle appearance and disappearance is that neutrons, protons and electrons have mass. Intuitively, mass seems substantial and indestructible; the material world of our experience is rooted in concrete matter – the stars, the Earth we stand on, our bodies, the objects around us. Although the *form* of matter may change, such as when a piece of wood is burned or a person dies, the total material content is usually imagined to remain the same. Surely all the atoms must go somewhere, they can't just disappear?

The decay of the neutron shows that matter can and does both disappear and appear on a subatomic scale. The mass of the neutron does not simply *change into* that of the proton and electron, for simple arithmetic shows that, at rest, a neutron weighs 0.78 MeV more than a proton and electron combined. During the transmutation, 1.4×10^{-30} kg of matter is destroyed. Where does it go?

A similar situation was discussed in Section 3.1 in connection with the masses of nuclei. The assembled nucleus is always lighter than the sum of the individual particles. The resolution comes from the theory of relativity, which tells us that mass is just a form of concentrated energy. A particle may gain or lose some mass by acquiring or discarding energy. In the case of the disappearing neutron, this particle's *entire* mass is converted to energy, some, though not all, of which goes to build the mass of proton and electron which suddenly appear. The remainder – the missing 1.4×10^{-30} kg – appears as energy of motion, exploding the decay products away from the centre of the transmutation at very high speed. Some energy also goes into the antineutrino which accompanies the electron and proton. Clearly a proper understanding of particle creation and annihilation requires the use of *relativistic* quantum theory, which takes proper account of the equivalence of mass and energy, and the very high speeds of the decay products.

In Chapter 2 Dirac's relativistic theory for fermions, among which are included the electron, proton and neutron, was described. If we wish to discuss the appearance and disappearance of such particles, we must expect to use this theory.

Recall that among the solutions to his equation, Dirac was puzzled by the existence of those with *negative energy*, which inevitably arise because of the additional space–time symmetry which is inherent in special relativity. Enigmatic though they are, the negative energy solutions cannot just be ignored. For a start, the negative energy levels extend downwards, without limit, in a sort of mirror image of the positive energy levels. This means that the usual ground state of, say, the hydrogen atom, is apparently not really a ground state at all but is poised over a bottomless well of negative energy states. There seems to be no reason why an electron in the hydrogen atom should not continue to drop down into lower and lower states, emitting more and more photons on the way. Indeed, as there is no lower bound at all to the system of quantum energy levels, it seems that all the matter in the universe must be inherently unstable, and descend into this negative energy pit among an escalating shower of radiation.

In 1930 Dirac suggested that to prevent this unhappy fate a rather radical idea should be invoked. He pointed out that, according to the Pauli exclusion principle, an electron would be prevented from making downward transitions from the usual ground state if all the troublesome negative energy levels are *already occupied* by other electrons. Of course, we see no sign of these negative energy electrons, even though an infinite number are needed to fill up all the levels. Dirac proposed that they are *invisible*. According to this picture, empty space – a vacuum – is not really empty at all, but is filled with an infinite sea of invisible negative energy particles. As the Pauli principle operates separately for different species of particles, they all require their own separate infinite seas.

In spite of its rather contrived aspect, this model of relativistic quantum matter has some real predictive power. Dirac argued that although downward transitions into the sea are forbidden by the Pauli principle, nothing prevents *upward* transitions of electrons *from* the sea into vacant *positive* energy levels. As a positive energy electron is 'normal' there is no reason why it should not be visible. It follows that such an upward transition would be seen by us as the sudden appearance of an electron in the universe where none had existed before. But this would not be all. If an electron in the sea is invisible, the hole left when the electron makes its upward jump must be visible, for the *absence* of a negative energy, invisible particle is equivalent to the *presence* of a positive energy, visible particle. Moreover, the same reasoning applies to the electric charge of the 'hole'. The absence of invisible negative charge is equivalent to the presence of visible positive charge. Thus the appearance of a newly created electron must be accompanied by the appearance of the 'hole' – a positively charged particle with the same mass as the electron (see Fig. 3.7).

Dirac's extraordinary idea implies that material particles can be created out of this infinite invisible reservoir, but only if they are accompanied by their 'mirror images', with the same mass, but opposite

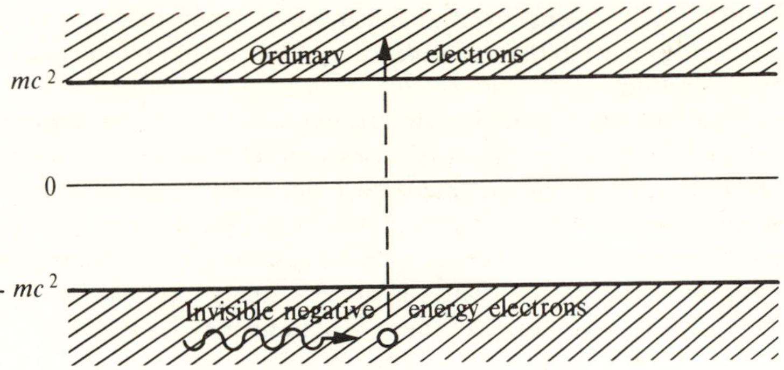

Fig. 3.7. Creation of matter and antimatter. Ordinary free electrons have positive mass-energies, greater than mc^2. Beneath these energies lies a gap of $2mc^2$ above an infinite sea of invisible negative energy electrons. If a gamma ray has enough energy to bridge the gap and raise one of these negative energy particles above mc^2, we see the creation of a new electron together with a 'hole' in the sea, which appears as a positron.

charge and, in fact, magnetic field and spin. Despite the extraordinary physical model on which the prediction of mirror particles is based, positive electrons were indeed discovered, in 1932, by the American physicist Carl David Anderson (Nobel prize 1936), whilst studying cosmic rays. Negative protons were found in 1955 using a particle accelerator at Berkeley in California, and mirror-neutrons in 1956. The name *positron* (symbol e^+) has been given to mirror-electrons, but the generic term is 'anti' rather than mirror. Thus, we speak of *antiprotons* ($\tilde{p}$) and *antineutrons* ($\tilde{n}$) and, in general, antimatter. There is no doubt that Dirac's bold prediction of antimatter on the basis of the mathematical unification of quantum theory and relativity is one of the great triumphs of theoretical physics.

An electron located in the negative energy sea will make an upward transition if it absorbs a photon, in the same way as upward transitions occur in atoms. The energy of the photon must be at least equal to the gap between the top of the sea (highest negative energy state) and the lowest vacant positive energy state. For free electrons of mass m this is $2mc^2$, which requires of the photon an energy of about 1 MeV, which puts it in the gamma ray region. To create proton–antiproton pairs requires gamma rays at least 1836 times more energetic, to supply the greater mass-energy of the proton. If more energy than the minimum is supplied by the gamma ray, the created particle–antiparticle pair will appear moving at speed, the extra energy being taken up by their motion.

As remarked, all subatomic processes may be reversed. The reverse of electron–positron creation occurs if an electron encounters a positron, when they will annihilate each other and emit gamma rays. This corresponds to a downward transition of an electron into a 'hole' (vacant energy level) in the negative energy sea, giving up the energy in the form of photons.

It can now be seen that the electromagnetic force can accomplish more than just bending the trajectories of electric particles: it can completely destroy or create them. Nor does the creation necessarily involve gamma rays. For example, if two protons are collided violently together, the electric field between them can create an electron–positron or proton–antiproton pair, using the energy of *motion* of the colliding particles. Symbolically these reactions may be written

$$p + p \rightarrow p + p + e^{+} + e^{-}$$

$$p + p \rightarrow p + p + \tilde{p} + p$$

which may be envisaged as follows. To create the particle–antiparticle pair, the colliding protons must have an energy of motion at least equal to the rest mass energy of the new particles. According to relativity the mass ($\times c^2$) of a moving particle is greater than its rest mass, and is in fact equal to its total energy – rest mass plus motion. Thus the moving protons are heavier than static ones. When they collide, this extra mass is stripped off and used to make new particles. To create a proton–antiproton pair, the colliding protons must obviously be at least twice as heavy as protons at rest, which means they must be moving faster than 86.6% of the speed of light. In this way, energy of motion is converted into new particles of matter.

The annihilation of electrons and positrons is not a straightforward affair. Like the proton and electron, the electron–positron pair attract each other electrically and can form a sort of mini-atom, revolving around their common centre of gravity. Known as *positronium*, this mini-atom has energy levels in the usual way, except the orbit with the smallest radius is no longer a true ground state, for positronium is unstable against decay by mutual annihilation. The size of positronium is twice that of the hydrogen atom, because the centre of gravity lies mid-way between the two equal mass particles rather than close to the very heavy proton. For the ground state, this size is about $\hbar^2/me^2$, a result which follows from the discussion on p. 71.

It might be thought that the two particles ought to physically touch each other in order to annihilate, but such a concept does not have the same meaning in quantum physics as in everyday life. According to the uncertainty principle,* a particle confined to a distance Δx has a momentum uncertainty $\hbar\ \Delta x$ which means that the particle cannot

* In the following we use the more accurate estimate $\hbar$ rather than h for the Heisenberg uncertainty (see p. 56).

remain at rest inside Δx, but must have some energy of motion. Until now, we have considered situations where the uncertainty energy is very much smaller than the rest mass energy of the particle, but in this section we wish to deal with relativistic situations. If Δx is small enough, the uncertainty energy will be so large that it is comparable with the rest mass. As explained in Section 1.5, when this happens, the relationship between momentum and energy is no longer the usual one for 'everyday' situations, but approaches that for the photon (Equation (1.12)): energy = momentum $\times\ c$, or $(\hbar/\Delta x) \times c$ in this case. For this to be equal to the rest mass energy mc^2, Δx must be about $\hbar/mc$, which for the proton is 0.21 fm and for the electron is 386 fm. The length unit is a characteristic length for each particle, and is a fundamental unit of relativistic quantum theory, known as the *Compton wavelength*, after the American physicist Arthur Holly Compton (1892–1962, Nobel physics laureate 1927). Its significance is that no meaning can be attached to the location of a particle inside a distance comparable with its Compton wavelength. The nearest one can get in quantum theory to the idea of two particles 'touching' is that they approach to within this basic distance. Sometimes this is visualized by imagining that an electron is not really a point particle, but is extended over a ball a few hundred fm across. This is, however, not strictly accurate. All one can say is that within this distance it is not possible to attach much meaning to the concept of 'an electron' as a well-defined particle with a definite mass.

In order for positronium to annihilate, it is necessary for the electron and positron to approach to within their Compton wavelength $\hbar/mc$, which is over 100 times smaller than the lowest orbital radius $\hbar^2/me^2$. However, quantum uncertainty in this orbital radius implies a finite probability for a sub-Compton encounter, which can be crudely estimated by taking the ratio of the Compton volume, $(\hbar/mc)^3$, to the orbital volume, $(\hbar^2/me^2)^3$, on the assumption that the particles move about randomly inside this region. Thus, the probability is about $(e^2 \hbar c)^3 \simeq (1/137)^3 = 3.9 \times 10^{-7}$, or one chance in about 2.6 million. (Notice how this probability is controlled solely by the so-called fine structure constant – the coupling strength of the electromagnetic field to matter – see p. 82.) However, it is not enough for the particles simply to reside this close: they must also emit a photon in order to annihilate.

In fact, it is necessary for at least *two* photons to be emitted in order for positronium to annihilate, for the following reason. When positronium is at rest in the laboratory it has no momentum. If only one photon were created it must move off at the speed of light, carrying momentum with it. As the creation of momentum is forbidden by the laws of mechanics (see p. 7), it is necessary for at least one other photon to be produced, to move off in the opposite direction and balance the books. (Recall that momentum is a vector quantity.) At this point the spins of the particles must be taken into account. The electron and positron have spin one half and the photons each have spin one. If the spin of the electron points in the opposite direction to that of the positron, their combined spins cancel to give a total of zero spin for positronium. On the other hand, if the spins line up, then positronium will have spin one. There are thus two varieties of positronium. In the former, annihilation into two photons can occur, each photon moving in an opposite direction to the other and spinning the opposite way. The total spin and momentum of the final photons is zero, as it was for the original positronium. In the latter case, there is no way the two photons can have total spin one, so *three* photons have to be emitted. This process is therefore considerably slower than the first. Experiments show that the two processes have lifetimes of 1.25×10^{-10} s and 1.40×10^{-7} s, respectively. Thus the annihilation of electron–positron pairs, whilst complicated and indirect, is nevertheless extremely rapid by human standards.

If antiprotons and antineutrons are present alongside positrons, and all are well isolated from ordinary matter, they could form into antiatoms and antimolecules. Indeed there could be regions of the universe in which planets, stars and even whole galaxies are composed of antimatter. These objects would be perfectly stable unless they encounter ordinary matter, when explosive mutual annihilation results. Thus, the solid, concrete materials of our surroundings are not as indestructible as we may imagine. In the presence of antimatter they would disappear totally in a burst of gamma rays.

With the benefit of these ideas we can now see that the two types of neutrino discussed in the previous section are indeed the antiparticles of each other as their names imply, for the *absorption* of an antineutrino in the inverse reaction

$$p + e^- + \tilde{\nu} \rightarrow n$$

is equivalent, according to Dirac's model, to the *emission* of a neutrino (see Fig. 3.8)

$$p + e^- \rightarrow n + \nu.$$

In fact, this equivalence extends to all particles involved in subatomic reactions, so that we can always remove a particle from one side of the reaction and replace it with its corresponding antiparticle on the other side. Thus we have possibilities such as

$$p \rightleftarrows n + \nu + e^+$$

$$p + \tilde{\nu} \rightleftarrows n + e^+$$

$$\tilde{n} \rightleftarrows \tilde{p} + \nu + e^+$$

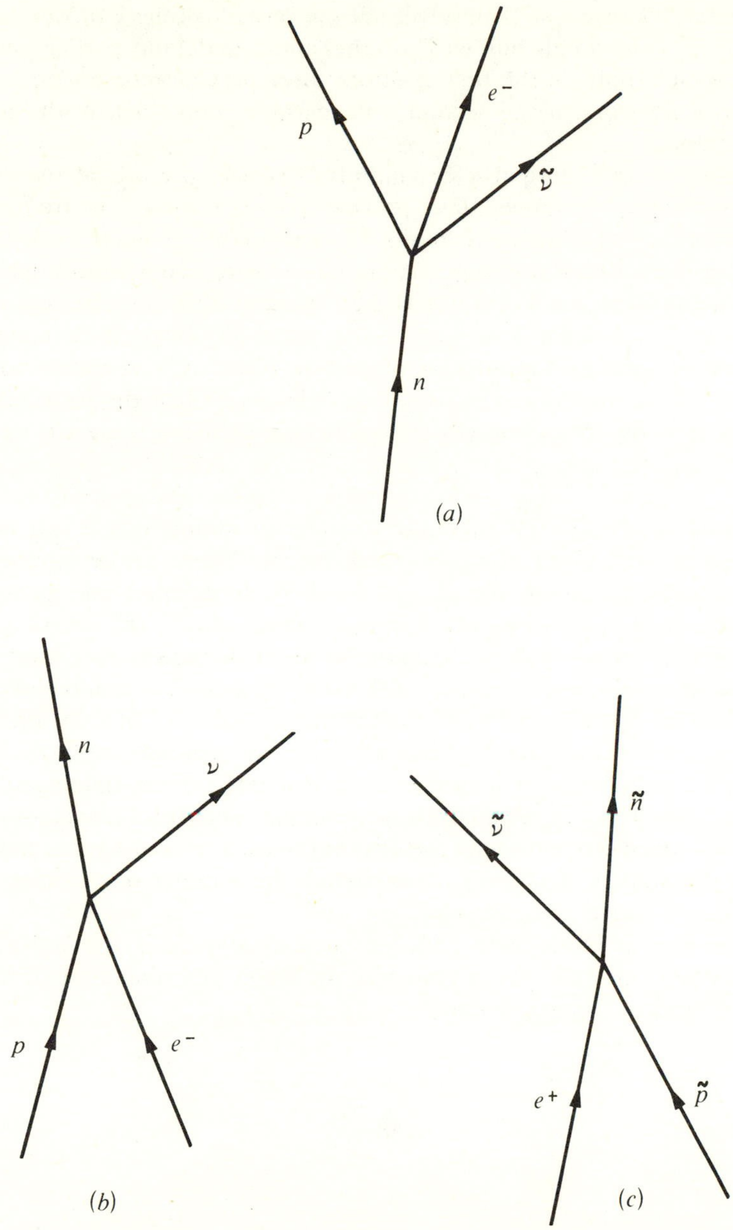

Fig. 3.8. Three faces of beta decay. (*a*) Disintegration of neutron, with the ejection of an electron; (*b*) electron capture by proton; (*c*) antiproton capture of positron. These diagrams, and others, may be derived from each other by interchange of the particle labels and/or reversing the arrows, remembering that the emission of a particle is equivalent to the absorption of an antiparticle.

As far as the *reactions* themselves are concerned, all these processes are equivalent, although our view of the initial and final particles is, of course, different. In the first example three particles come together to make a neutron, in the second, two particles make a neutron and a neutrino.

Using particle–antiparticle symmetry, we can predict all the other reactions shown above. The process $p \rightarrow n + \nu + e^+$ is frequently observed as an alternative mode of beta decay in neutron-deficient nuclei. On p. 102 it was explained how lead with 116 neutrons decayed by the process $p + e^- \rightarrow n + \nu$, i.e. by capturing an electron from outside the nucleus. Now we see the possibility of *positron* emission rather than electron absorption. This type of decay is very common: it occurs in the radioactive isotopes of light nuclei that do not bind the inner electron orbits very closely because of their lower electric charge (see Fig. 3.6*c*).

In spite of the success of Dirac's model of antimatter, it can never constitute a thoroughgoing physical theory. There are a number of reasons for this. First, the idea of an invisible negative energy sea of particles can at best be only a device for visualizing the creation and annihilation mechanism. It cannot be taken seriously as a model of quantum interactions. Secondly, the same arguments cannot be invoked for bosons (particles with an integral number of units of spin) because they are not subject to the Pauli principle. As we shall see in the next chapter, there exist in nature bosons that possess associated antiparticles. Thirdly a deeper analysis of quantum mechanics reveals serious mathematical problems with the attempt to treat matter wave equations as a description of *single* particles when the number of particles of a given type might be changing.

The answer to all these criticisms was discovered in the 1930s and developed into a full blown theory in the 1940s and thereafter. It forms the subject of the next chapter.

4

The new forces

Gravity and electromagnetic forces have been familiar to people since time immemorial. By contrast, the nuclear forces, called weak and strong, were only discovered when twentieth-century technology enabled physicists to use subatomic probes to examine the internal structure of atoms and their nuclei. The effects of nuclear forces are now well known. They keep the sun and stars alight, operate bombs and power stations, and illuminate the hands of watches.

Nuclear forces are most conspicuous, quite naturally, inside the nucleus, and in the previous chapter something of nuclear structure and stability under the action of these forces was described. The nucleus is, however, a complicated place, and our understanding of both the weak and strong forces has gained much from the study of the interaction between individual particles. In many cases the interactions are controlled by manipulating beams of subatomic particles in enormous electrical accelerating machines, some of which extend for several kilometres, and allowing them to collide with other beams or fixed targets. Examination of the scattered primary particles and their debris enables physicists to piece together a detailed picture of the nature of nuclear forces.

The search for an understanding of these forces in the 1930s and 1940s soon brought with it the discovery of new forms of matter – previously unknown subatomic particles that participate in the weak and strong interactions. The advance in understanding of mathematical theory, particularly the quantum theory of fields and the union of quantum theory and special relativity, showed that the nature of subatomic forces is inseparable from the structure of subatomic matter. Whereas the classical forces of gravity and electromagnetism appear to operate *between* matter, the weak and strong forces (and, indeed, the electromagnetic force when it displays its quantum features) are found to be closely interwoven with the particles that they affect. We shall see that not only are the internal properties of the particles themselves, such as mass, magnetic field, electric field and so on, dependent on the forces of nature, but the very forces are communicated by still more particles that are subject to more forces, and so on.

The subatomic world has no clear division into passive particles moving under the influence of active forces. Instead, we see a complex network of activity, with particles and forces closely involved in each other's affairs. The key to this micro-universe is the marriage of quantum theory, field theory and relativity, three of the most difficult topics of modern physics. The language of the marriage is mathematics, of a form well beyond the level of this book. Nevertheless, the underlying physical ideas are accessible to the reader who has got this far, though some of the description will necessarily involve using analogies and simplified models.

4.1 *The quantum theory of fields*

The discussion given in Chapter 3 on the creation and annihilation of material particles, such as electrons and positrons, highlighted the central role of special relativity in subatomic affairs. Important though $E = mc^2$ is for the conversion of energy into the mass of the created electrons or positrons, it does not of itself explain how this creation comes about, or how other particles (e.g. the neutron) can suddenly disappear.

A clue to the appearance and disappearance of particles was available from the earliest days of quantum theory: the photon. In Chapter 2, processes were discussed in which photons were created or destroyed by the motions of electric particles, the most obvious being from the transitions of electrons between atomic energy levels. We are happy to accept the fact that the number of photons in the universe is not fixed once and for all, but is continually changing as processes occur within matter. Whenever a domestic light is switched on, new photons appear, only to disappear again as they are absorbed into the walls of the room. On a cosmic scale too, the number of photons is steadily increasing as the stars pour out radiation into the vast emptiness of space. All this seems quite unremarkable.

One reason that makes it easy to accept the fact that photons come and go is that their creation has a familiar and readily visualizable classical limit. Imagine the moving electrons becoming more and more numerous until their individuality is lost and they build up into an electric current. Similarly, as billions of photons are piled together, quantum fluctuations and uncertainties become relatively less and less important compared to the huge total energy of all the photons. Wavelike aspects dominate over particle-like aspects as we approach the limit of a classical electromagnetic wave, possessing a definite wavelength, amplitude and phase. The emission of photons *en masse* can then be identified with the familiar process of the generation of electromagnetic waves by flowing electricity: for example, radio waves from an aerial.

As mentioned in Section 2.7 the existence of a classical wave limit is closely associated with the fact that photons are *bosons* and do not obey the Pauli exclusion principle. There is no corresponding classical limit for electrons, protons, neutrons or neutrinos, for they are all *fermions*. In spite of this it seems that these particles can appear and disappear *individually*, even though we cannot have anything like the generation of a macroscopic electron or neutrino wave by a macroscopic source.

This fundamental difference between photons and the fermion particles gives the photon a somewhat separate status. When quantum theory is applied to the electromagnetic field it describes new particle-like properties, but when it is applied to, say, electrons, it describes new wavelike properties. In short, quantization 'particle-izes' waves and 'wave-izes' particles. Thus it establishes common features between such dissimilar animals as electromagnetic waves and electrons. However, they are still different in many respects. It is hard to envisage a photon as a 'real' particle on the same footing as an electron. What we now want to do is to close the conceptual gap that still divides the photon from particles like electrons, so that we may do for the latter what can be done for the former, namely, describe creation and annihilation.

To map out the necessary route to this proposed unification, let us re-examine what a photon really is. Imagine for a moment an electromagnetic wave travelling through space like a wave travelling along an infinite string. In quantizing the wave we require a description of how energy resides in the vibration and can be extracted or inserted in discrete amounts. We are not directly concerned with the actual motion of the wave. Consequently, it is easier conceptually to imagine a *standing* wave. In the case of the string this can be produced by stretching the string tightly, clamping the ends and plucking it.

Standing waves were described in Section 2.6 in connection with matter waves. We can also produce them for electromagnetic waves by enclosing the system inside a metal box, whose walls reflect the waves perfectly. The wave configuration inside the box will be rather like the disturbance of air inside an organ pipe – a fixed wave pattern. Recalling the discussion of standing waves given in Section 2.6, we note that only certain allowed frequencies of vibration can 'fit' into the box or on the string. However, if the box is very large and the wavelength very small, then the allowed frequencies are so closely spaced that we do not really notice any restriction. This is certainly the case in practice: light has a wavelength of about 10^{-7} m, whereas a laboratory box might be, say, 1 m in size. This means that some 10^7 waves 'fit' into the box, and the increase in frequency caused by fitting in one more wave is minute.

Focus attention on one particular allowed frequency, or mode of vibration as it is called. According to classical physics the *strength* of the vibration can be anything at all. Now imagine an excited atom in the box. It can emit some energy into the field and make a downward transition as a consequence. How does this energy reside in the field? It must be as an *excitation* of one of the field modes. Which mode is determined by the amount of energy released, through the formula $E = h\nu$, which then fixes the frequency ν. But Planck's relation tells us that only this fixed quantity of energy, $h\nu$, or a multiple thereof, can be released (or absorbed) by the atom. It follows that a field mode can only contain definite fixed units of excitation energy. If it contains one unit, we say that one photon is present with frequency ν; for two units, there are two photons, etc. We know from the Pauli principle that there is no upper limit to the number of units, or quanta, of energy in each mode. When the number is very large the total energy is too great for us to notice by comparison the tiny increments of $h\nu$. This corresponds to the classical limit which allows continuous transfers of energy to the field.

Viewed this way, the photon does not seem very much like a particle. However, a closer examination shows that the differences are not too great. An energetic material particle such as an electron, when confined to a box, will bounce backwards and forwards between the walls. The limitations of the Heisenberg principle will completely prevent us from locating the position of the particle in the box so long as we require it to have a fixed momentum and energy. Thus, we can think of the particle probability wave as spread throughout the box in the same way as the electromagnetic wave. Moreover, the photon excitation contains a fixed quantity of energy, momentum and spin just as the material particle does.

According to this picture, the creation of photons is to be viewed like the plucking of a guitar – as a sudden increment in the excitation of one of the modes of vibration. To encompass this feature in the description of all subatomic particles it is necessary to extend the idea of matter waves – to place them on an equal footing with electromagnetic waves. Instead of regarding matter waves as describing the behaviour of a single subatomic particle, we treat them as forming a general *matter field* and regard a particle such as an electron as a *quantum of excitation of a mode of the matter field*, after the fashion of the photon. For example, an electron is a quantum of the electron–positron field, a proton a quantum of the proton–antiproton field, and so on. From now on we shall think of 'particles' and 'field quanta' as one and the same thing. In this way, it is as though quantum theory has been applied twice to matter: once to produce wavelike field behaviour, then a second time to recover some particle aspects from quanta of the field.

Using this 'second quantization', as it is known, it has been possible to build a complete relativistic theory of interacting quantum fields capable of describing a wide range of subatomic particle activity. Whilst the theory contains some unsatisfactory features, and many physicists believe it is still in a tentative condition, its success gives confidence that the essential ideas are basically correct. Of course, at the level of this book we can only give a very superficial glimpse of the workings of this powerful theory.

Easily the most successful application of the full theory of second-quantized relativistic fields is to a system of interacting electrons, positrons and photons, a subject known as quantum electrodynamics. Many detailed predictions of the theory have been checked to phenomenal accuracy and appear to give a completely satisfactory account of nature.

Quantum field theory requires a new way of thinking about microphysical processes. Something of the novel ideas, in the language of which much of the later material in this book is based, will now be described. Because the treatment is necessarily heuristic, we shall frequently appeal to diagrams for explanation. While this mode of analysis is commonplace even among professional physicists, it carries with it the usual dangers of pictorial representation. The reader should always remember that the diagrams are partly schematic and are not supposed to be pictures of the real microworld. For the purposes of illustrating the relationship between different but connected processes, they are invaluable. But they are not intended to be a substitute for detailed mathematics.

The most primitive process in electrodynamics is the emission of a photon by an atom. Physically it is easy to visualize, and its consequences are familiar in daily life. A pictorial representation of this process may be constructed by extending the idea of a space–time map, which was discussed in Section 1.5. The history of the atom which emits the photon is represented by a solid line, and the photon by a wavy line. Fig. 4.1 depicts this event, which shows the atom making a transition from an excited state to the ground state as a consequence, and the photon shooting off to the right. The atom recoils slightly to the left.

Other processes are shown in Fig. 4.2. In (*a*) an electron collides with a positron and annihilation occurs in the form of two photons (see p. 108). In (*b*) the reverse process of pair creation by a gamma ray photon occurs. The diagram for this process involves, unlike (*a*), only one photon. In order to conserve momentum in this lopsided arrangement it is necessary for an additional system (such as another charged particle, not shown here) to take up some of the recoil. Diagrams of this type were used in Section 3.3 in connection with the beta decay of the neutron.

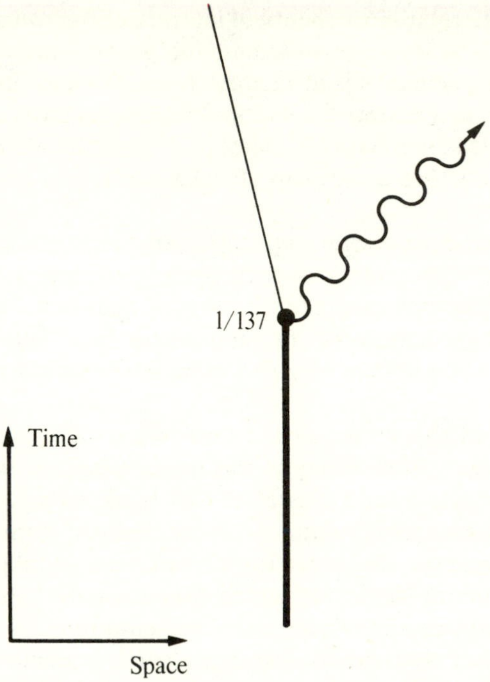

Fig. 4.1. Photon emission. This space–time map shows an excited atom (heavy line) emitting a photon to the right and recoiling to the left. The likelihood of the process is regulated by the number 1/137.

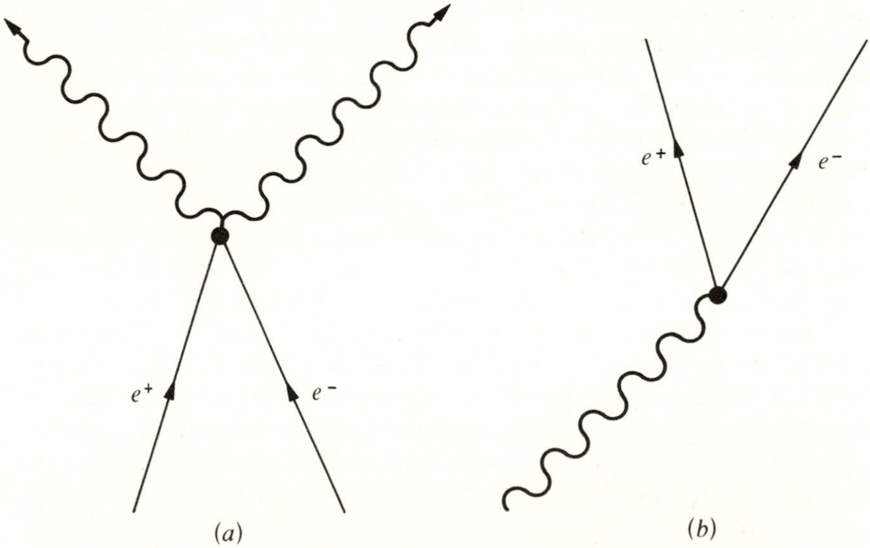

Fig. 4.2. (*a*) Pair (electron–positron) annihilation into two gamma ray photons. (*b*) Pair creation by a gamma ray photon.

One feature, which will be typical of all such diagrams, is that the photon emission or absorption is shown as a single event, occurring at one space–time point, marked on the diagrams with a little dot. This point is called a vertex, to be envisaged as the place and time where, in Fig. 4.1, the photon suddenly appears and the atom is suddenly in its ground state. Now this is a difficult thing to reconcile with the classical picture. In the pre-quantum model of light emission, one pictured a smooth transition from one electron orbit to another resulting in the slow radiation of a light wave, with the wave troughs and crests gradually building up into an extended undulation. To get a clearer picture we must remember the probabilistic nature of quantum mechanics. For example, Fig. 4.1 is only one of an infinite number of diagrams in which the photon appears somewhere on the atom's world line (see p. 71). To calculate the probability of a photon being present after a given time, we must take into account *all* the diagrams which show an emission somewhere before that time. If we consider millions of excited atoms together, the combined, averaged effect of all their statistically regulated emission events would indeed approach the characteristics of the smooth, steady emission displayed in the classical picture.

Let us try to focus on what is really happening in these photon emission and absorption processes, according to the picture by quantum field theory. It is helpful to return to the analogy of a vibrating guitar string to represent modes of the quantum field. Suppose we have two strings on a guitar of exactly equal length, mass and tension. It is a phenomenon well known to musicians that if one of the strings is plucked, the other will gradually start to vibrate in sympathy (see Fig. 4.3). The explanation for this is that the two strings, by belonging to the same guitar, are not really independent mechanical systems, but are coupled together through the material of the bridge. The coupling,

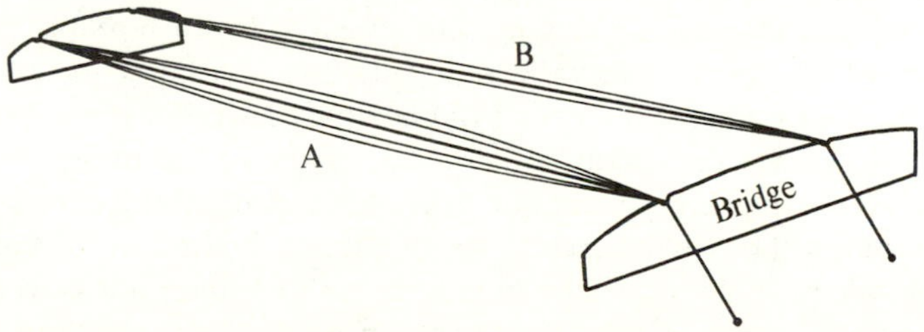

Fig. 4.3. Sympathetic vibration. String A is plucked, but B also vibrates a little if it is tuned to the same note (even an octave lower). This resonance phenomenon is caused by the coupling through the bridge, and produces a transfer of the wave excitation energy from A to B. A similar phenomenon in subatomic waves causes the creation of particles such as photons as excitations in the wave fields.

although weak, communicates the vibrations from the plucked string to the unplucked one, thereby causing it to resonate. Because the second string is tuned to vibrate at exactly the same frequency as the first, it will be synchronized with it. This means that every time a little push comes through the bridge, the second string will be moving in the correct way to absorb the available energy from the wood. Consequently, more and more energy builds up in the second string – the amplitude of its vibration grows progressively greater. This energy is, of course, supplied by the plucked string. Thus, the resonance amounts to a *transfer of energy* between the strings. Furthermore the *rate* at which the energy is transferred depends on the strength of the coupling. If the strings are a long way apart, or if the bridge is very stiff, then not much vibration will get through.

To make the analogy with wave fields somewhat closer, it is better to think of very long strings vibrating, not in their lowest tone, but in a very high harmonic, as shown in Fig. 4.4. Under these circumstances, there will also be the possibility of stimulating in the second string a vibration which is one octave lower; this is well known, middle C in one octave can resonate the note C in another octave. On the string this corresponds to a wavelength twice as long (i.e. half the number of waves fit on to the same length of string). In fact, the first tone will in general stimulate a little bit of *all* available lower octaves and superimpose them on the original tone of the plucked string. If the coupling between the strings is efficient, these lower harmonics will be quite strongly encouraged.

Consider now how this picture is modified by quantum theory. A real guitar is far too large for us to notice quantum effects, so think of an atomic sized one. Recall the discussion of Section 2.6. When the first string is plucked it cannot vibrate with *any* particular amplitude or energy: only integral multiples of $h\nu$ are allowed according to Planck's original hypothesis. Suppose only one quantum of excitation energy $h\nu$ resides in the string after plucking. Obviously it is not possible for this quantum of energy to *gradually* pass into the second string by resonance, because we are not allowed to have only part of one quantum on each string. Either the quantum is on one string or the other. We can, however, say that there is a definite *probability* that, after a certain time, the quantum will be on the second string but not the first. If the coupling is strong, the probability will be high after a short time; if it is weak, we must wait a long time before finding the quantum in the second string. The actual transfer of vibration between the strings can be envisaged as a sudden jump. First one string is vibrating and the other not, then vice versa. The suddenness may appear strange, but it is not really an observable effect, for all we can measure is that, at a certain moment,

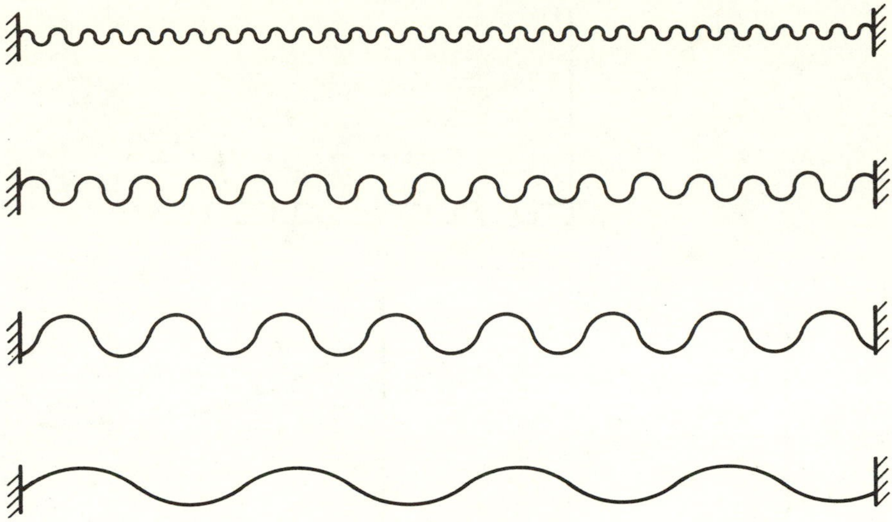

Fig. 4.4. The high harmonic A can induce sympathetic vibration of lower frequencies 1/2, 1/4, 1/8, . . . in B, C and D, respectively.

the second string contains the energy. We cannot measure exactly when the jump happened, nor can we calculate it. The Heisenberg uncertainty principle forbids it. All that the theory contains are probabilities.

If our quantum has a high frequency it is possible for it to divide itself in half and stimulate *two* quanta in the next lower octave. The energy exactly balances because the lower octave has half the frequency: $h\nu = h(\frac{1}{2}\nu) + h(\frac{1}{2}\nu)$. In this way, a quantum of excitation energy in one string has been destroyed and two created in the other string. This binary creation is less probable than the creation of a single quantum of the same frequency ν. By coupling to still lower octaves, four, eight, sixteen . . . quanta of progressively less energy may be created instead, but this is far less probable.

Before returning to the discussion of the creation and annihilation of particles in wave fields, it is instructive to examine the classical limit as the guitar gets bigger and bigger. If just one quantum of excitation is present, we have seen that it must jump discontinuously between the strings. However, if many quanta of excitation are present (the string is vibrating vigorously in its energy level) then it is possible for both strings to be excited simultaneously, because a multiplicity of quanta are available to share between them. The effect of the coupling is to cause the quanta to jump across, usually one at a time, from the highly excited string to the other one. Fig. 4.5*b* shows how the excitation energy of the

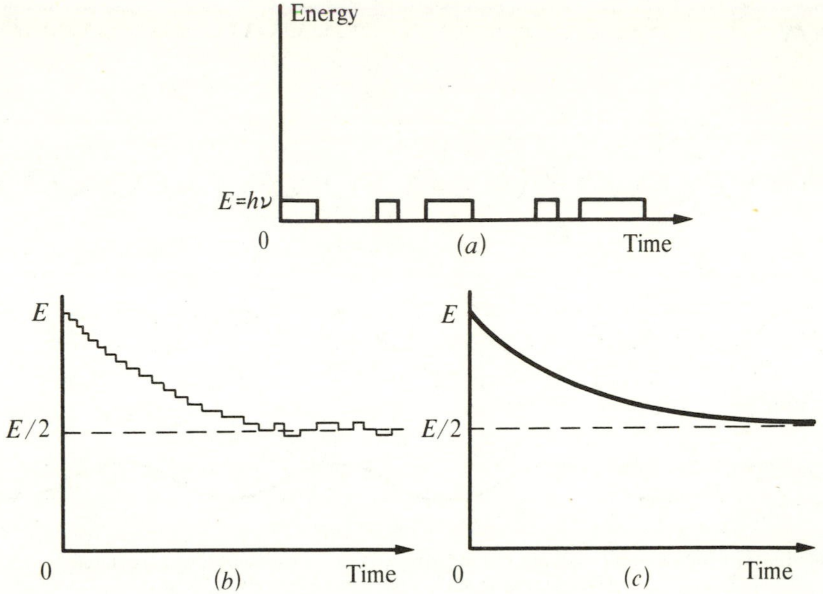

Fig. 4.5. Energy transfer between guitar strings (losses neglected). (*a*) If the system shown in Fig. 4.3 contains just one quantum of excitation energy, $E = h\nu$, then the initially excited string will share it on a 'shift' basis with the other. The graph shows the moments when the energy is on the plucked string. (*b*) The plucked string contains many quanta, which leak to the other string, usually one at a time. Rapidly at first, then levelling out, about half the quanta eventually transfer, after which random fluctuations occur. (*c*) In a realistic guitar, the quanta are so numerous (billions of billions) that we do not notice the sudden jumps, and the decay in energy appears smooth.

first string slowly leaks away in jumps. The weaker the coupling the slower the decay rate. On a real vibrating guitar string, there are billions and billions of quanta, and we do not notice the little jumps (*c*). Random statistical fluctuations become less and less important relative to the total energy on the string, so the slightly non-uniform rate of jumps evident in (*b*) is not apparent in (*c*). There appears to be a smooth, continuous flow of energy out of the plucked string into the other.

One final point. In a real guitar, the energy is not conserved in the strings. Much of it leaks away into the fabric of the guitar where it is dissipated as heat. In the discussion here we have ignored this effect.

The analogy between guitar strings and subatomic particles is closer than their very different classical nature suggests. The creation and disappearance of a quantum of excitation on a string is exactly like the creation and disappearance of a particle. Remember, in quantum field theory, a particle *is* an excitation of the corresponding matter wave field mode. The essential feature is the *coupling* between different wave fields (guitar strings). In Figs. 4.1 and 4.2 this coupling is electromagnetic, which is rather weak (remember $e^2/\hbar c \simeq 1/137$ which is a small number). The role of the guitar strings is now played by the electromagnetic field (photons) and the electron–positron field, or the matter waves of an electron in an atom. The transfer of quanta between the strings corresponds to the emission and absorption of photons.

The analogy is clearest in Fig. 4.2*b* where a photon creates an electron–positron pair. The incoming photon is like one quantum on the first string which suddenly transfers its energy to create an excitation on the second string (electron–positron field), i.e. the photon disappears and an electron–positron pair appears. The reverse process of pair annihilation, shown in Fig. 4.2*a*, is an example of lower octave resonance, where two (or three) quanta – photons – are created instead of one. Because the coupling is small (1/137) the probability of these many-quanta processes are smaller the greater the number of quanta involved (recall the lifetime data quoted on p. 110 for this process).

It is important to remember that the diagrams only represent *one* possible process. We could have other multi-quantum processes such as five-quantum annihilation. To calculate the exact probability of electron–positron destruction we should have to calculate *all* possible higher quantum processes and combine together all their probabilities. Fortunately, however, nature has been kind, and given us a very weak electromagnetic coupling. Each extra photon emitted is roughly 1/137 times less probable than the previous process, so in practice we only need calculate the probabilities of the first few processes. These considerations are sometimes expressed by writing 1/137 by every photon vertex to remind us that every time another photon is created or destroyed (emitted or absorbed) the probability of the total process gets about 1/137 times lower (see Fig. 4.1).

In the case of the atom making a transition between energy levels and emitting a photon, we could if we wanted envisage the atom being destroyed and another created at the vertex. However, in this case (unlike the pair annihilation) we have available a first quantized picture – without the need for matter appearance and disappearance described by second-quantized field theory. So we can think of the excited states of the atom as closely similar to the excited levels of the guitar string, as we did in Section 2.6. The photon, of course, must still be regarded as an excitation of the electromagnetic wave field.

In the above analogy we assumed all along that the guitar strings A and B had been properly tuned. It is instructive to consider what happens if the unplucked string B is out of tune; too slack, or too long, for example. We know what happens according to the classical picture:

as the strings cannot resonate, the energy transfer between them is inhibited, however good the coupling. If the mismatch is not too great, the plucked string A may still elicit a partial response from B. The transmitted vibrations try to drive B at A's frequency, but out-of-tune B does not respond very naturally and the effect is small. The greater the mismatch, the lower the response. How can we view this situation quantum mechanically?

According to the Heisenberg uncertainty principle, Equation (2.8), B can 'borrow' energy from A, provided it pays it back again after a very short time. The bigger the debt the shorter the duration of the loan. For a small mismatch between the strings, the energy of a one-quantum excitation of string B is nearly the same as that of A, so the debt is small and the loan long: there is a good probability that we will find the excitation energy in string B. For a big mismatch, the jump is greater, and the excitation can spend less time in string B, so the probability of encountering it there is small. The name given to a 'not-quite-the-right-energy' quantum is *virtual*. Virtual quanta have a limited lifetime. To distinguish them, we call quanta with an unlimited life (correct energy) *real*. Obviously some virtual quanta are 'almost real' because they have nearly the right energy and live a very long time.

A fascinating consequence of the existence of virtual quanta emerges when we go to quantum field theory, for we now have the possibility of *virtual particles*. For example, in order to create an electron–positron pair, a gamma ray photon needs an energy at least equal to their combined rest mass (about 1 MeV). However, if the photon has less energy than this, *virtual pair creation* can still occur. What this means is that the photon suddenly turns into an electron–positron pair, which then annihilates a brief moment later and creates another photon. Because it all happens in a microscopic region of space the particles are also allowed to suspend the law of conservation of momentum (recall $\Delta x \Delta p \simeq h$), so the annihilation process only re-creates a single photon, rather than the two or more that must appear after the annihilation of a *real* electron–positron pair.

Fig. 4.6 shows the diagram of this process. Its existence has a profound and far-reaching consequence for the nature of subatomic matter, for it implies that an ordinary photon is really being continually destroyed and reborn. Indeed, a high-energy photon spends a lot of its time as an electron–positron pair, because the energy debt is small. Low-energy photons are less likely to be found in this form. The existence of virtual charged particle pairs associated with photons causes a curious phenomenon. If two photons encounter each other they will in all probability pass straight through one another – photons do not couple directly because they do not carry electric charge. (We ignore

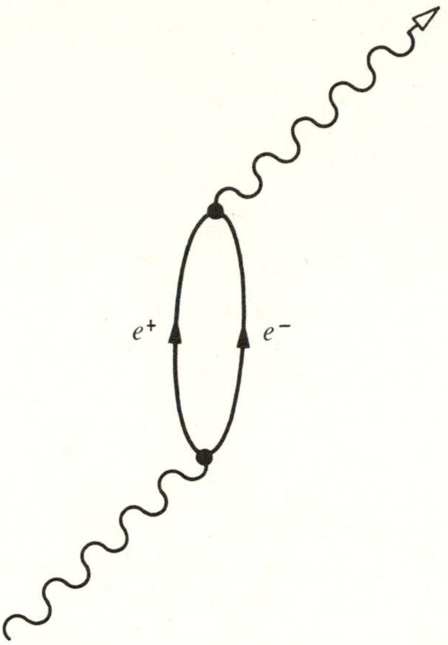

Fig. 4.6. Transmutation of the photon. The Heisenberg principle allows a photon to borrow enough energy to convert briefly into a virtual electron–positron pair, which quickly annihilates to give back the photon again.

here their minute gravitational coupling which undoubtedly exists.) However, if both photons happen to have converted to the electron–positron form at the moment of encounter, then these virtual electric particles *will* interact, with the result that one photon will effectively act on another. The result is photon–photon scattering (see Fig. 4.7). The fact that light can exert a force on light is a peculiarly quantum effect; unfortunately it is just too small to have been detected in the laboratory yet.

Just as the photon carries with it a cloud of virtual charged particles, so a free electron or a positron carries with it a cloud of virtual photons. For a brief while an electron can emit a photon, so long as it is reabsorbed again quickly. This process is shown in Fig. 4.8. Of course, the long-wavelength photons can last longer because they carry less energy. In general, a virtual photon can live for a time

$$\Delta t \simeq h/\Delta E \simeq \nu^{-1}$$

where we have used the fact that the borrowed energy ΔE is just that of the photon, $h\nu$. During this time, the photon can travel a distance $\frac{1}{2}(c/\nu)$ and back, which is equal to one half its wavelength. This is in contrast with a real photon, which can detach itself completely from an electron and travel off to a great distance, enjoying an independent existence. In this way, the virtual photons describe the *nearby* field of an electric

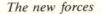

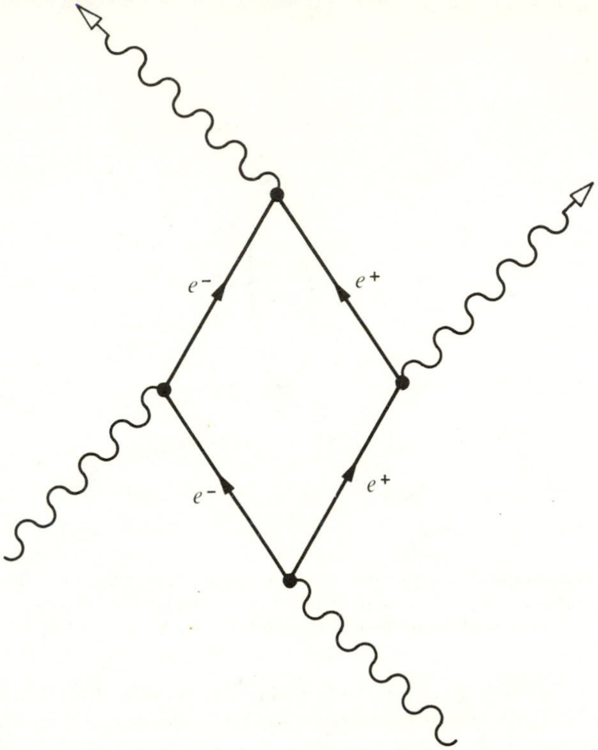

Fig. 4.7. Light–light scattering. When virtual electron–positron production occurs in the fashion shown here, two photons couple together, producing photon–photon scattering (two in, two out).

charge whereas the real photons belong to the distant (radiation) field. Virtual photons cannot rob the electron of energy permanently, but real photons do. We can therefore identify virtual photons with the electrostatic field, which falls off as the inverse square ($1/r^2$) of the distance r from the charged particle, whereas the real photons represent the electromagnetic radiation that falls off more slowly ($1/r$) and hence travels to great distances. The electrostatic field contains energy but does not let it flow away. On the other hand, radiation (real photons) does transport energy away from the charge.

In classical electrodynamics one charged particle may act on another in two rather different ways. First, the electric and magnetic forces between them can act directly from one to the other: the electric repulsion in a nucleus is one example. These forces are short-ranged. Secondly, one particle may emit radiation which travels through space and disturbs the other. These forces are long-ranged. This separation

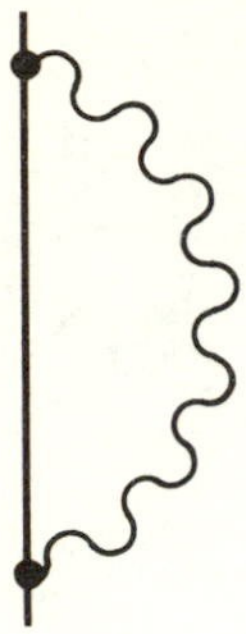

Fig. 4.8. All charged particles are surrounded by a cloud of *virtual* photons like this one. They can be thought of as responsible for producing the nearby electric field of the particle. Any radiation emitted is due to *real* photons.

into two different modes of action is not an idea consistent with the theory of relativity, because only for one type of observer are the electric forces static. Another observer who is moving relative to the first observer would not regard the electric charge as static. According to the rules of relativity, no action can propagate faster than light, so even the short-ranged forces must really be communicated between charged particles through the intermediary of the electromagnetic field.

With the help of our fully relativistic quantum field theory we can see how this works. When one electron moves close to another, feels a force and is turned away, we can describe this as the *transfer of a virtual photon* between them. The arrangement is depicted in Fig. 4.9. The electron on the left emits a virtual photon and suffers a recoil. Instead of being reabsorbed by the same electron, as in Fig. 4.8, the photon is absorbed by the other electron, which also recoils. Taking into account the effect of all such virtual photon exchanges (including the transfer of two, three, . . . photons) gives an explanation of the reason why one electron experiences a force and is repelled by the other.

Because of the energy–time uncertainty relation, it is not possible to tell which particle has emitted and which has absorbed the virtual photon. The order of such closely separated events cannot be distinguished. Consequently we often draw the photon line horizontal. Because of this it may be imagined that quantum processes violate the rules of relativity theory, which demand a precise order of cause and effect, and only permit the transfer of information at a finite speed. If the time order of the emission and absorption of virtual quanta cannot be measured does it not imply something similar to instantaneous transfer of information between the particles? Could we not use virtual photons to send real messages faster than light? The answer is no. Nature has arranged for quantum theory and relativity to co-exist with

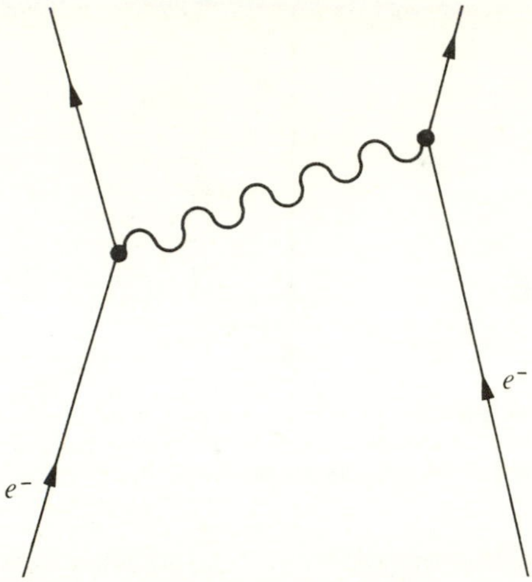

Fig. 4.9. The force of repulsion between two charged particles such as electrons may be computed from the effect of the transfer of virtual photons between them.

wonderful consistency. To convey information using electromagnetic signals we need at least one whole wave to encode it. However, we have seen that virtual photons cannot travel beyond a distance of one wavelength, so they are useless for messages. Only real photons can carry information. Apparent causality violation by virtual photons is a Pyrrhic victory against relativity, because in the absence of information transfer, the difference between cause and effect is transcended.

The virtual photons which bustle around an electron must be envisaged as closely tied to it by the uncertainty principle, never venturing very far away. If, however, the electron were to suddenly disappear, the virtual photons would have nothing to hold them, and would fly off as real photons. Such a sudden disappearance of the electron is not as improbable as it sounds. We know that should it encounter a positron it will indeed be annihilated. So we can think of the gamma rays that are produced by electron–positron annihilation as the residue of the virtual photons which are always carried along by these particles, and which are liberated when the electron and positron disappear.

By building up diagrams of the type shown in Fig. 4.7–4.9, all known electrodynamic phenomena can be accurately described, and their relative probabilities calculated. More examples are shown in Fig. 4.10. In (*a*) an electron absorbs an incoming real photon for a short time and then emits another: the electron recoils under the impact of the incoming photon and this acceleration causes it to radiate again. The action of the electromagnetic field on the electron thereby causes a reaction of the electron on the field. If the incoming photon is very

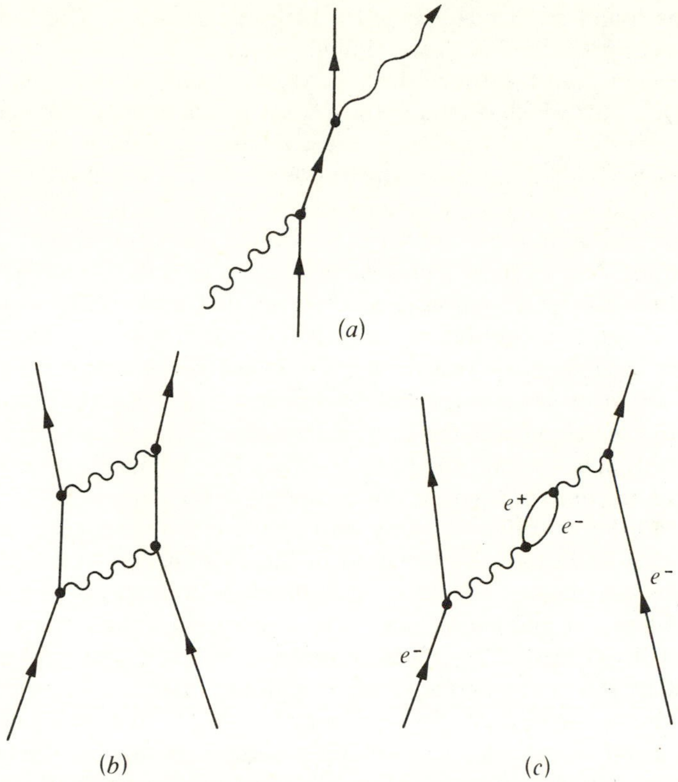

Fig. 4.10. (*a*) Compton scattering. An energetic photon (short wavelength) collides with an electron and transfers some energy. (*b*) Electron scattering from a two-photon exchange. (*c*) Higher-order process in which the virtual photon being exchanged between two scattering electrons itself decomposes momentarily into an electron–positron pair.

energetic, the recoiling electron may be left with considerable energy at the expense of the outgoing photon, whose frequency and energy are diminished relative to the first photon. This process is well known in the laboratory and called Compton scattering, being the phenomenon which earned Arthur Compton his Nobel prize.

When we take into account higher-order, less probable, processes with many vertices in the diagrams, things become very complicated and actual numerical calculations long and tedious. Only the smallness of the coupling (1/137) saves us from having to compute many of these multi-vertex processes. Diagrams of the type used here are called Feynman diagrams after the American physicist Richard Phillips Feynman (Nobel physics prize 1965), who first used them as an aid to computation in quantum electrodynamics.

The reader should not imagine that the presence of a cloud of virtual photons round an electron is just a heuristic gimmick. These photons cause real, measurable effects, though because of the smallness of the coupling constant the effects are always small. One of the most famous is the tiny shift which occurs in the energy levels of all atoms, but most conspicuously for hydrogen. It comes about because the presence of the electrically charged nucleus disturbs the virtual photon cloud surrounding the orbiting electron, changing its energy by a very small amount.

There are a number of ways in which the energy shift can be understood. One of these pictures the electron as suffering a slight recoil every time a virtual photon is emitted, so that as it orbits around the proton in the hydrogen atom its path takes on a sort of jittery aspect. The tiny excursions towards the nucleus lower the energy of the electron somewhat, because of its greater proximity to the attractive proton. The outward excursions raise the energy. However, because the electric field of the proton is not uniform, but falls off as $1/r^2$, the gain in energy from the outgoing excursions does not compensate the loss from the ingoing ones, and the net effect is a tiny shift of the electron's energy level.

Another important confirmation of the existence of virtual photons concerns the magnetic field of the electron (and the positron). The effect of the virtual cloud here is to alter very slightly the magnetic moment by 0.1159652%. It is possible to measure this tiny quantity extremely accurately, and agreement with the calculated prediction for the magnetic moment has been obtained to better than one part in 10^9 There is no doubt that this stunning success is one of the greatest triumphs of modern theoretical physics, and gives great confidence that the underlying ideas of relativistic quantum field theory are correct.

4.2 *The strong interaction: a first look*

Armed with the techniques and concepts of the theory of quantum fields, great progress can be made in understanding the interactions between subatomic matter which are not of the electromagnetic variety. In Section 3.1 some features of the strong nuclear force, which supplies the cohesive binding of the nucleus, were discussed. This force, which operates between protons and neutrons, has been studied directly by scattering these particles off each other and observing the disturbances on their motion, and indirectly by examining the properties of systems that are bound together by the nuclear force; namely the nuclei.

It is known from these observations that the nuclear force is very strong and short-ranged; outside 1 fm it dies away rapidly. A great deal of information can be obtained by studying the simplest composite nuclei. One of these is deuterium, a stable isotope of hydrogen consisting of a single proton stuck to a single neutron to form a so-called deuteron. Deuterium is found naturally among ordinary hydrogen (with which it is chemically identical), for example in water, where the deuterium component, because of its extra weight, is sometimes called heavy water. The deuteron is only very weakly bound, and neither the union of two protons, nor of two neutrons, can occur in isolation.

Another simple nucleus is that of tritium, a still heavier isotope of hydrogen containing a proton and two neutrons. The tritium nucleus is occasionally, somewhat unconvincingly, called the triton. It is beta-radioactively unstable, and decays by electron emission, with a half-life of 12.3 years, into a closely related nucleus – a stable isotope of helium with two protons but only one neutron, a sort of mirror of the triton. Because two mirror nuclei differ only by the interchange of a proton and a neutron, they provide important confirmation of the *equality* of the forces between neutron–proton, proton–proton and neutron–neutron pairs, once the electric repulsion between the protons has been allowed for.

A further important property of nuclear forces, already mentioned on p. 90, is that the nuclear material is relatively stiff: the bits of nucleus stick together strongly, but do not squeeze the nucleons together into a tight ball. It was supposed that this effect is due to the fact that only nearest-neighbour nucleons attract each other, so that the forces *saturate* and do not grow with increasing numbers of particles.

Any explanation of the strong nuclear force must be able to account for these elementary properties. The first attempt, a brilliant piece of deductive guesswork, was made by the Japanese physicist Hideki Yukawa in 1935, on the basis of quantum field theory. The essential concept is easily understood in analogy with the electromagnetic force. In the previous section it was explained how the exchange of a virtual photon leads to a force between electrically charged particles. Yukawa proposed that each proton and neutron is surrounded by a new type of field to which it couples through a nuclear 'charge' in the same way that an electron couples to the electromagnetic field though its electric charge. This new field will have quanta of excitations – virtual particles – in analogy to photons, and the exchange of these virtual particles between neutrons and protons leads to a force of attraction.

Yukawa's new quantum field could not be too closely analogous to the electromagnetic example, or he would end up reproducing the features of that force. Electric action is long-ranged, and repulsive between similar particles. Yukawa needed a short-ranged attractive force. It will now be explained how these differences can be achieved.

First consider the range of the force. If it is to operate by the exchange of virtual quanta, then the interaction between two particles cannot take place beyond the limits to which the exchanged quanta are allowed to travel. These limits are dictated by the uncertainty principle which says a quantum of energy E can only live for a time $\Delta t \simeq \hbar/E$, and hence can travel at most a distance $c\Delta t \simeq \hbar c/E$ before being absorbed. In the electromagnetic case, the energy can be made as low as we please by considering virtual photons of sufficiently low frequency ν. This is why the electromagnetic force is long-ranged: the energy of interaction falls off as $1/r$ and the force as $1/r^2$. To obtain a short-ranged force, Yukawa proposed that the quanta of the new nuclear force field should have a *rest mass*. Thus, the energy E that has to be borrowed to create a virtual quantum now has a lower bound: it must be at least equal to mc^2, for a particle of mass m. The uncertainty principle then allows it to exist for only about $\Delta t \simeq \hbar/mc^2$ after which it must be absorbed again. During this time light can travel a distance $\hbar/mc$, and because relativity forbids the particle to travel faster than light the maximum range of the force is about $\hbar/mc$ (the Compton wavelength, in fact–see p. 109). By making m large, the range can be made as small as one pleases. To obtain a range of 1 or 2 fm m works out at about 300 electron masses – mid-way between the mass of the electron and proton. For this reason, the quanta of the nuclear field are called *mesons*.

Turning now to the second difference from electromagnetism, Yukawa's virtual quanta are supposed to attract similar particles in the nucleus. It can be shown from the full mathematical theory of quantum fields that the *spin* of the virtual quanta that are exchanged determines whether the force is attractive or repulsive. The photon, with spin one, repels like charges. Gravity, with spin two, attracts. The simplest choice for attraction is spin zero: this was Yukawa's choice.

Of course, Yukawa did not base his theory on a heuristic argument using the uncertainty principle. He developed a proper mathematical theory of the meson field that surrounds the protons and neutrons, and received the 1949 Nobel physics prize. From this theory he was able to give the exact form of the nuclear force field surrounding a proton or a neutron, in the same way that Maxwell's theory of electrodynamics allows one to calculate the force or energy surrounding a point electric charge. Yukawa's theory gave an interaction energy which falls off very rapidly with distance in an exponential way: e^{-kr}/r. Fig. 4.11 shows a graph of this expression, and its form is that of a bottomless well with a steep side that rapidly approaches zero, under the influence of the exponential factor, outside a certain distance labelled 'range'. The constant k is equal to $mc/\hbar$, which is the typical distance travelled by a virtual particle that we calculated above from the Heisenberg principle. As m approaches zero, the exponential factor becomes one, and Yukawa's expression approaches the form of the electrostatic energy, $1/r$.

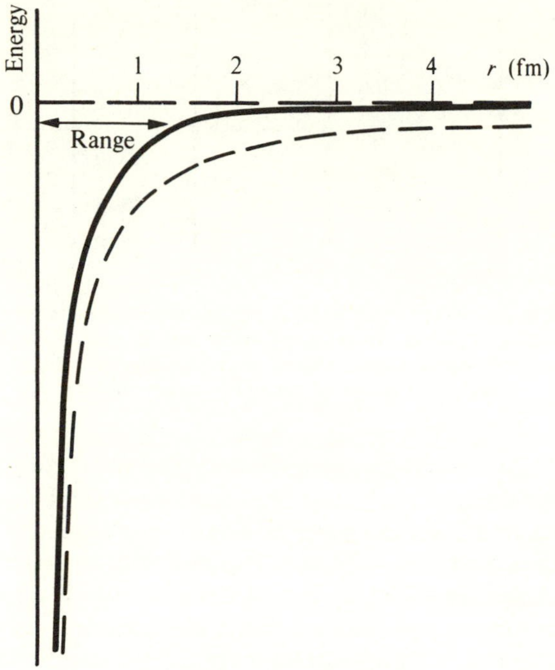

Fig. 4.11. The nuclear force field. Yukawa's meson field is short-ranged. Its attraction energy falls rapidly to zero outside about 1 fm, in contrast with the electrostatic force (broken curve), which is long-ranged.

The actual strength of the force between two nuclear particles is determined by the nuclear 'charge' on each of them just as the electric charge determines the electric force. The nuclear charge cannot be calculated from Yukawa's theory but must be measured by experiment. It turns out to be 45 times larger than the fundamental unit e of electric charge. This means that the nuclear force between two protons is $(45)^2$ or about 2000 times stronger than the electric force. If we call the nuclear charge on a proton or neutron g, then the quantity $g^2/\hbar c$ is the analogue of the fine structure contrast. One finds that $g^2/\hbar c \simeq 15$, in contrast with 1/137 for the electromagnetic coupling. For this reason the nuclear force is called the *strong interaction*.

Yukawa's picture of nuclear forces can be given a diagrammatic representation (see Fig. 4.12) showing a meson being exchanged between two nuclear particles. There will also be contributions from the exchange of two, three, four, . . . mesons. The strength of these contributions changes in the ratio $g^2/\hbar c$, $(g^2/\hbar c)^2$, $(g^2/\hbar c)^3$, . . . but because the nuclear coupling is so strong, these numbers are all comparable with

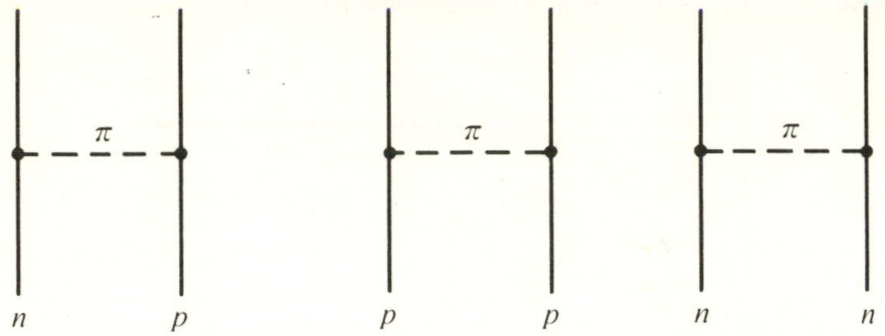

Fig. 4.12. Origin of the nuclear force. Yukawa's explanation was the exchange of a virtual meson, labelled π, between two protons and/or neutrons, in direct analogy to the virtual photon exchange which causes the electromagnetic force (shown in Fig. 4.9). The π lines have been drawn horizontally for convenience.

each other. Unlike electrodynamics, where the small coupling (1/137) usually enables the effects of multi-photon processes to be ignored, here we cannot regard meson exchange as a small perturbation. This means that calculations involving the strong interaction are much harder to perform, and the results of approximations in which all but a small number of mesons are ignored are much less reliable.

The restriction to nearest-neighbour forces can also be incorporated into Yukawa's picture, to explain the saturation property of the strong interaction. Recall from Section 3.3 that a neutron may 'turn into' a proton by emission of an electron and antineutrino. Similarly, we may suppose a neutron may become a proton, and vice versa, by emitting a meson. This could lead to the situation shown in Fig. 4.13, where a virtual meson is swapped between a neutron and a proton. The meson would have to be electrically charged so that charge is conserved at each vertex. Either a negatively charged meson travels left to right or a positive meson right to left. The overall effect is to *swap the identities* of the neutron and proton. For this reason the force which results is referred to as an *exchange force*. It is obviously a private arrangement between two nucleons, so involves only one neighbour, thus explaining the saturation property, and the rigidity of the nuclear material. Direct evidence for the existence of an exchange process comes from the study of the scattering between neutrons and protons, when protons are sometimes observed where neutrons 'should' be.

Focus attention now on a solitary neutron. It can still create virtual mesons, but in the absence of a companion it must reabsorb them itself before too long. On average they only travel about half the radius of the nuclear force – about 0.7 fm. The neutron is thus surrounded by a cloud of massive virtual particles, and although the neutron is overall neutral some of the mesons will be electrically charged (see Fig. 4.14). If

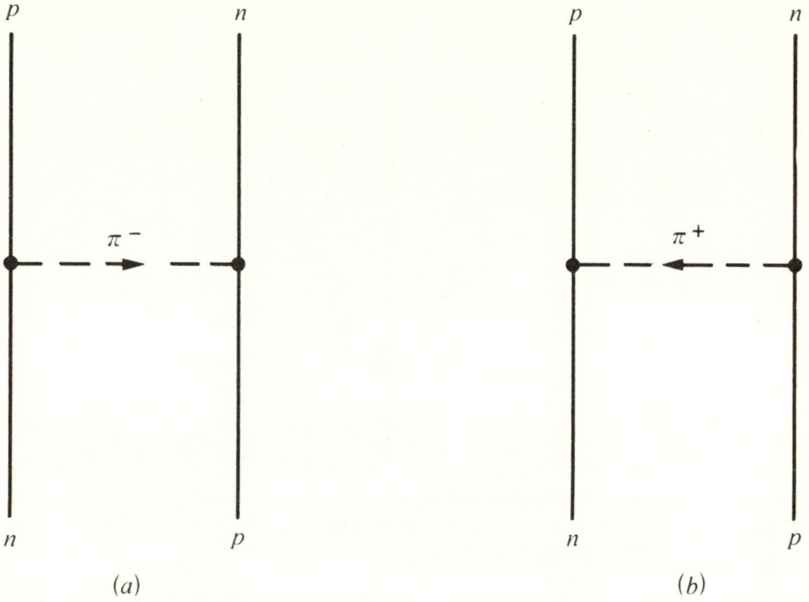

Fig. 4.13. Exchange forces. Neutrons and protons may swap identity by meson exchange.

another nuclear particle enters this cloud, it will interact strongly with the virtual mesons and experience the strong force. In contrast, a neutrino, an electron or a photon will not feel the nuclear force of the cloud. However, both the electron and the photon will feel the *electric charges* in the virtual meson cloud, even though the neutron itself is electrically neutral.

The presence of a cloud of virtual charged mesons around a neutron is rather like an electric current distribution. In both cases there is overall electric neutrality because there are equal numbers of both positive and negative charges moving about. All this implies that in spite of its neutrality, the neutron can still have some electric action. What evidence do we have of the existence of the virtual meson cloud?

In 1933 it came as something of a surprise when an experiment performed by the German physicist Otto Stern (1888–1969, Nobel physics prize 1943) showed that the neutron is a tiny magnet. It was known that the neutron has a spin of one half like the electron, but in the absence of electric charge, this spin cannot itself cause magnetism. It is found that the magnetic field surrounding a neutron is just like that which would be caused by a spinning point particle with a negative charge of about 1.9 times the fundamental unit *e*. It is compelling to ascribe this magnetism to the motions of the charges in the neutron's meson cloud.

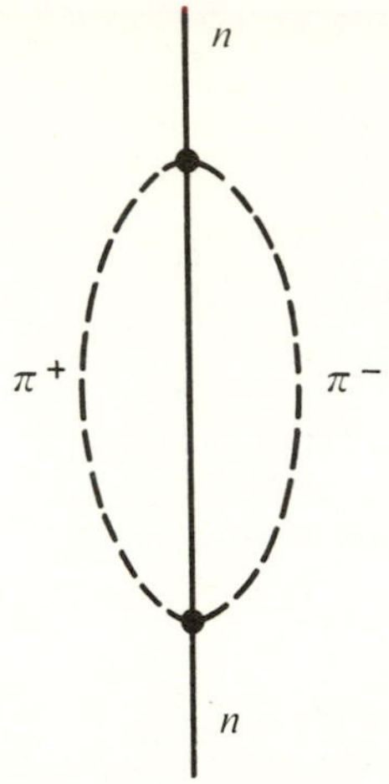

Fig. 4.14. The presence of a cloud of virtual charged mesons dressing a neutron endows it with electric action. Overall neutrality is achieved by having an equal number of positive and negative charges present. Compare with Fig. 4.8.

The charged meson cloud has also been probed by scattering electrons off neutrons. These projectiles are found to feel electric forces from the cloud, and the distribution of charge inside it can be mapped out in some detail.

A similar picture applies to protons. The presence of a meson cloud gives the 'dressed' proton a charge distribution that can also be mapped out by electron scattering. The proton magnetic moment turns out to be more than 2.8 times larger than it should have been were it merely a more massive version of the positron (which has no cloud of charged mesons around it) even though they both have the same quantity of positive charge and the same spin. In fact, this is a good example of the tight interplay between the forces of nature and the structure of matter. The positron is subject to the electromagnetic force, and has a magnetic moment which can be calculated from Dirac's equation plus a tiny correction due to the virtual photon cloud which dresses it (see p. 130). The proton appears at first identical apart from mass. However, being subject to the strong force as well drastically changes its magnetic properties, introducing an enormous correction to the Dirac value.

In the electromagnetic field, virtual photons mediate interactions between charged particles and surround solitary particles in a cloud. If the parent particle is shaken violently enough, some of the virtual photons can be shaken off and become real photons. Classically this corresponds to the production of radiation by accelerating charged particles. Quantum mechanically, it is necessary to supply the virtual quanta with sufficient energy to overcome the imprisonment of the uncertainty principle, so that they can escape. In the same way it should be possible to knock out *real* mesons from the virtual cloud around nuclear particles, if enough energy can be provided in the knock to supply at least the meson rest mass. Remember that the virtual mesons are living on borrowed energy. To promote them to real mesons requires the debt to be cleared completely, a service which can be supplied by the incoming projectile.

One place to look for liberated mesons is where two nuclear particles strike each other at speed. In the 1930s machines could not easily achieve the modest energy requirements, which entails accelerating protons to half the speed of light. Instead, the search began among cosmic ray processes. Early in 1937 a new charged particle was duly discovered among cosmic ray products at Pasadena in California by Anderson and his colleague Seth Neddermeyer. With a mass of 206.8 electron masses it was at first assumed to be Yukawa's meson. However, disappointment soon followed when it was realized that the newly discovered particle did not interact strongly with nuclear matter, so could not be a quantum of the nuclear force field. The 1937 particle is now known as the *mu meson*, or *muon* for short (written μ) and will be discussed further in the next section.

The world had to wait until 1946 before the Yukawa meson was finally identified among cosmic ray products in a photographic emulsion by the British physicist Cecil Frank Powell of Bristol University and his colleagues of the Ilford photographic company. These mesons are called *pi mesons*, or *pions* for short, and written π, as anticipated in Figs. 4.12 and 4.13. As expected there are three of them: an electrically neutral one π^0, with a mass of 264.1 electron masses, a positively charged one, π^+, with a mass of 273.1 electron masses, and a negatively charged one, π^-, also with a mass of 273.1 electron masses. In fact the charge ones, $\pi^\pm$ are antiparticles of each other. Also as expected, all three pions have zero spin. They are *bosons*. For his discovery Powell received the 1950 Nobel physics prize. In spite of the success of Yukawa's theory, it is known to be only part of a very complicated nuclear interaction.

Pions are produced in high-energy collisions between nucleons, for example

$$p + p \rightarrow p + p + \pi^+ + \pi^-$$

which can be achieved by accelerating a beam of protons in an electrical machine and smashing them into a target. They are also produced when nucleons are annihilated. The reason for this is easy to visualize. Recall that the neutron is surrounded by a cloud of virtual pions; the proton, because it is also electrically charged, has both a cloud of virtual pions and a cloud of virtual photons. Should a proton suddenly disappear, these virtual particles are liberated and real pions and photons appear. Thus proton–antiproton annihilation produces pions and photons:

$$p + \tilde{p} \rightarrow \pi^+ + \pi^- + \pi^0$$

or

$$\rightarrow 3\pi^0$$

or

$$\rightarrow 2\gamma.$$

Sometimes more pions or photons than this are produced, if the proton and antiproton can collide with sufficient energy for conversion into the extra particles. Similarly, pions are produced by neutron–antineutron annihilation

$$n + \tilde{n} \rightarrow \pi^+ + \pi^- + \pi^0$$

or

$$\rightarrow 3\pi^0.$$

The discovery of the three pions marked one of the turning points of twentieth-century physics. Not only was it realized that matter could exist in hitherto unknown forms, but the quantum field basis of the strong interaction was confirmed. Evidently new physics could be unlocked by breaking matter apart under high-energy bombardment, opening up the prospect of more new physics at still higher energies. In addition to this, pions are found to display startling new behaviour: they are *unstable*. Electrons, protons, photons and neutrinos are all apparently completely stable – they have unlimited lifetime and do not transmute into anything else when left alone. Even neutrons are only weakly unstable, and can easily be stabilized inside nuclei. In contrast, pions only live for a fleeting moment. In only the briefest fraction of a second they disintegrate into more familiar particles, which is why they are not observed among ordinary matter. The existence of highly unstable pions suggested that there might be other types of matter – still more particles – whose existence had been overlooked because of their extremely short lifetimes.

4.3 *The weak interaction: a first look*

In spite of the great success of Yukawa's theory, it cannot explain all nuclear interactions. Beta decay, for example, cannot be controlled by the strong force, because it involves electrons and neutrinos, which do not participate in strong interactions. Historically, a theory of beta decay predated Yukawa's work by a year. In 1934 Fermi constructed a theory of beta decay which was also based on analogy with the electromagnetic interaction. Processes such as neutron decay can only be explained by introducing yet another new force of nature to drive the disintegration. By nuclear standards, the 15-minute lifetime of the neutron is extremely long, indicating a very weak process at work. Accordingly, the fourth known force of nature is called the weak interaction, in contrast with the strong interaction.

Rather than discuss the decay of the neutron

$$n \rightarrow p + e^- + \bar{\nu}$$

let us concentrate instead on the more symmetric process

$$n + \nu \rightarrow p + e^-.$$

(Remember that the first process is essentially the same as the second if we regard the absorption of a neutrino by the neutron as equivalent to the emission of an antineutrino, on the basis of Dirac's hole theory that the presence of a particle is equivalent to the absence of its antiparticle.) Now that quantum field theory has taught us not to be afraid of interactions that create and destroy particles, this process does not seem quite so strange. We could depict it by a diagram of the form shown in Fig. 4.15, where a neutron and a neutrino enter a highly localized

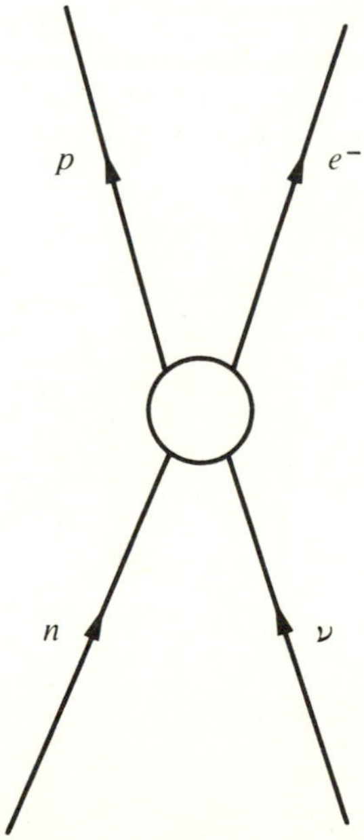

Fig. 4.15. Schematic representation of the beta decay process $n + \nu \rightarrow p + e^-$ where the blob represents some unknown interaction process. Note that this is the reverse of the process shown in Fig. 3.8*b*.

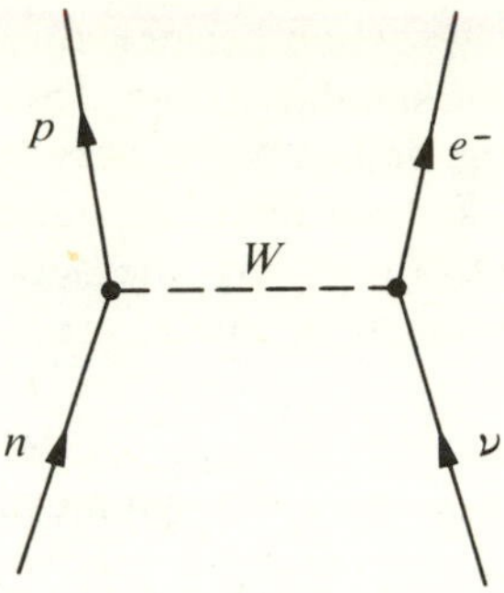

Fig. 4.16. Fermi theory of the weak force. In analogy to Figs. 4.9 and 4.12 the weak force is envisaged here as due to the exchange of a new virtual quantum, labelled *W*. To conserve electricity at each vertex *W* must be charged.

interaction region, and emerge as a proton and an electron. Analogy with the electromagnetic (and Yukawa) interaction suggests that this occurs by the exchange of a new type of messenger particle (virtual field quantum) labelled *W* (see Figs. 4.16 and 4.17). Clearly *W* must carry electric charge (W^+ or W^-), because the interaction carries negative electricity from left to right, or positive from right to left.

We could regard the interaction as causing the destruction of a neutron and the simultaneous creation of a proton at the left-hand vertex, and the similar disappearance of a neutrino and appearance of an electron at the right-hand vertex. Alternatively we could think of the neutron as 'turning into' a proton and a W^- particle, followed by the absorption of W^- by the neutrino, which converts it into an electron. Of course, the *W* could also be envisaged as passing in the reverse direction

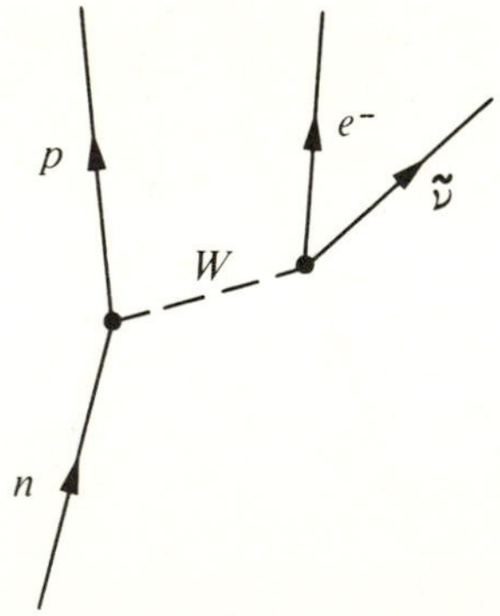

Fig. 4.17. This diagram is really equivalent to Fig. 4.16 (just change the outgoing $\tilde{\nu}$ to an incoming ν). It describes the beta decay of the neutron, and is a magnification of the process depicted in Fig. 3.8*a*.

(as a W^+). All these modes of description are essentially equivalent (W is only a *virtual* quantum anyway).

To understand these remarkable transmutations, let us pursue the analogy with electrodynamics more closely (compare with Fig. 4.9). When two electrons repel each other by exchanging a virtual photon one can think of each moving electron as a tiny electric current, and regard this current as responsible for producing (and absorbing) the photon. In the same way we regard the neutron–proton line as a sort of current – not electric, but a new type of current, a *weak* current (after 'weak' interaction). That is, the neutron and proton should be envisaged as 'charged' with a new 'weakness' quality. We can think of the weak charge as passed on to the proton by the neutron during the transmutation at the left-hand vertex. Similarly the neutrino carries a weak charge and passes it on to the electron. Fig. 4.16 therefore describes a current–current interaction closely similar to the electromagnetic case. Naturally the antiparticles ($\bar{p}, \bar{n}, e^+, \bar{\nu}$) carry weak charge of the opposite sign. Other particles, such as photons, do not have any weak charge.

We see that weak charges and currents are closely similar to the electric charges and currents. The main difference is that weak charge does not exert any macroscopic forces. The reason for this is the very short range of the weak interaction (which may even be zero). The range is, of course, governed by the mass of the intermediate particle W. What experimental evidence is there concerning the range? The most direct evidence would be the actual detection of a real (as opposed to virtual) W particle in the same way that Yukawa's pion was eventually detected. A measurement of its mass would then yield the necessary information about the range. However, as yet no W has been found, probably because its mass is too large for the energies currently available in our accelerators.

It is possible to place an upper limit on the range of the weak force by studying the process discussed on p. 102

$$p + e^- \rightarrow n + \nu$$

which occurs when a neutron-deficient nucleus captures an orbital electron. If the weak force of the proton extended beyond the periphery of the nucleus, it would reach out to the orbiting electrons and produce the above reaction much more readily than is observed. So the range must be less than 1 fm. A direct search for W fails to find anything with a mass less than about 20 000 MeV (more than 20 proton masses) so this confines the range of the weak force to less than 10^{-2} fm. If W exists, quantum field theory predicts it to have a spin of one unit, like the photon.

There is one other respect in which the mechanism for the weak interaction differs fundamentally from the electromagnetic interaction, and indeed, the gravitational and Yukawa interactions too. All the latter forces can operate between just two particles of matter which can be either fermions or bosons; for example, electron–proton, or pion–neutron. In contrast, the weak interaction described here involves four distinct particles, all of which are fermions. Processes such as $n \rightarrow p + e^- + \bar{\nu}$ and $n + \nu \rightarrow p + e^-$ are clearly four-fermion interactions, involving one or more fermions *turning into* other fermions. We shall return to this point later.

Using a four-fermion theory of the kind described above, based by analogy on electrodynamics, Fermi was able to account for the known features of various weak interaction processes. In due course Fermi's theory required some modification (see Section 5.4) but the essential conceptual structure remains the same.

The classical forces of nature, gravity and electromagnetism, manifest themselves at the macroscopic level by their action on the *motion* of particles of matter. Even the strong nuclear force is readily envisaged as something restraining the protons in a nucleus from exploding away from each other. In contrast, the decay of a neutron, $n \rightarrow p + e^- + \bar{\nu}$, does not at first seem to be an example of a force at all. However, with the benefit of relativistic quantum field theory and diagrams like Figs. 4.16 and 4.17, we can see that it *is* a force, whose field quanta (W) have not yet been observed in isolation. This provides another illustration of how the forces of nature are inextricably wound up with the structure of matter, for the weak force here acts not on the motion of matter, but on its *form*. It converts neutrons into protons and neutrinos into electrons. Clearly we cannot study the forces of nature without also studying the many forms of matter which arise in the guise of a variety of subatomic particles. The physics of forces and of particles are one and the same thing.

The discovery of the muon in 1937 added a new dimension to weak interaction theory, for μ is found to decay with a half-life of 2.2×10^{-6} s into an electron and two neutrinos. There are two muons, one with negative electric charge and its antiparticle with a positive charge. The decay schemes are

$$\mu^- \rightarrow e^- + \nu + \bar{\nu}$$

$$\mu^+ \rightarrow e^+ + \nu + \bar{\nu}$$

and from the fact that the decay products consist of an odd number of fermions, we may deduce that the muon is also a fermion; it has spin one half, like the electron, rather than spin zero, like the pion, a fact which may be confirmed experimentally by studying the bound states of the μ^+e^- system.

The presence in these decays of weakly interacting particles such as electrons and neutrinos strongly suggests that the force responsible for disintegrating muons is the same weak force that effects the demise of neutrons. In any case, the electromagnetic force could not be responsible otherwise a gamma ray photon would be produced instead of neutrinos. Also, as remarked, muons do not participate in the strong interaction. Moreover, the muon decay is also a four-fermion process, like beta decay. In fact, there are genuinely four different fermions involved, because experiments show that the two neutrinos, ν and $\bar{\nu}$, are *not* antiparticles of each other. They are two *different kinds* of neutrino. One type is always associated with electrons, and is denoted ν_e; the other type is associated with muons and is denoted ν_μ. Thus, muon decay should really be written

$$\mu^- \rightarrow e^- + \bar{\nu}_e + \nu_\mu$$

The Fermi model of weak interactions can be applied to this process, or the equivalent and more symmetrical one

$$\mu^- + \nu_e \rightarrow e^- + \nu_\mu$$

in which the muon changes into a muon-neutrino and emits a negatively charged W^-, which converts the electron–neutrino into an electron (see Fig. 4.18). Here the μ^-, ν_μ pair play the role of the n,p pair in beta decay (compare $n + \nu_e \rightarrow e^- + p$).

We must envisage the muon and its neutrino as carrying a weak charge too, and the μ^-,ν_μ line as a weak current which interacts through the W with the ν, e^- current. The distinct nature of muon- and electron-neutrinos really obliges us to regard the muon-weak charge as distinct from the electron-weak charge (both are apparently conserved). However, as will be explained below, the muon-weak current behaves in all respects like the electron-weak current, so it is perhaps better to think of them as two *branches* of a *universal* interaction.

In the last section it was mentioned that pions are also unstable, and disintegrate after a very short time. In fact, charged pions decay into muons with a half-life of 2.6×10^{-8} s:

$$\pi^+ \rightarrow \mu^+ + \nu_\mu$$

$$\pi^- \rightarrow \mu^- + \bar{\nu}_\mu.$$

Each muon is produced with its associated muon-neutrino.

As the weak force is apparently responsible for exploding neutrons and muons, it might be expected that it would also account for the decay of pions, especially as both the decay products are known to be weakly interacting. However, the universal Fermi model requires a

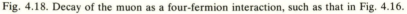

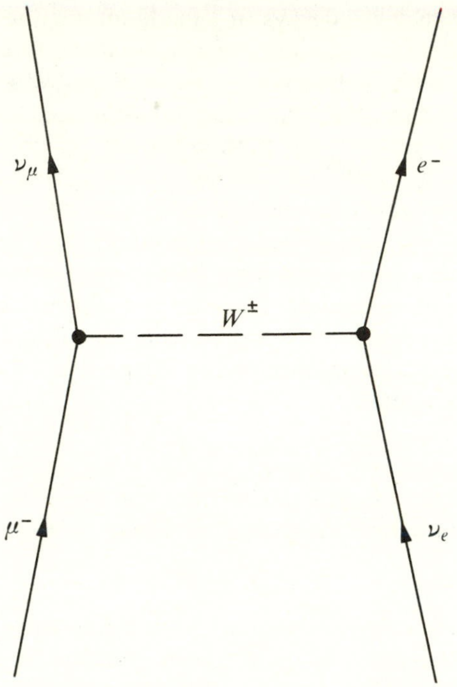

Fig. 4.18. Decay of the muon as a four-fermion interaction, such as that in Fig. 4.16.

four-fermion interaction, whereas pion decay involves two fermions and a boson.

Nevertheless, it is possible to account for this process using the universal model, when the possibility of *virtual* particles is taken into account. For an exceedingly brief time (about 10^{-23} s) the pion can convert itself into two nucleons via the strong nuclear force. Although the mass of the nucleons greatly exceeds that of the pion, the necessary mass-energy can be borrowed briefly using the Heisenberg uncertainty principle. In fact, just as a photon must be regarded as spending some of its time as an electron–positron pair, so we must regard all pions as existing part of the time in the form of a nucleon–antinucleon pair. Thus, the pion is not to be imagined as remaining inert until it decays. Instead, it is continually disintegrating into nucleons and then reforming again in a restless sequence of activity

$$\pi^+ \rightarrow \underset{\text{(virtual)}}{p + \tilde{n}} \rightarrow \pi^+$$

$$\pi^- \rightarrow \underset{\text{(virtual)}}{\tilde{p} + n} \rightarrow \pi^- .$$

Now so long as the pion remains a boson, it is safe from destruction via the weak interaction. However, during its short vacations as a pair of *fermions*, it is vulnerable to the four-fermion process. This is when the weak interaction grabs it. We must imagine the $p,\tilde{n}$ pair, which is the other face of the π^+, as being wooed by two competing forces: the strong interaction which wants to coalesce them into a π^+ once again, and the weak interaction which wants to turn them into a muon and a neutrino. Because the strong force is so much stronger than the weak one it nearly always wins. Billions of times the pion is safely reconstituted. But eventually, by the laws of quantum probability, the weak interaction will let one go. That is all it needs, for once it has operated to produce a μ^+,ν_μ pair, the competition from the strong interaction suddenly and irreversibly disappears, because the μ^+,ν_μ particles do not participate in the strong force. Thus, the muon and neutrino are left unhindered to fly away as real particles. The sequence may thus be represented as a two-stage process:

$$\pi^+ \underset{\text{strong}}{\longrightarrow} \underset{\text{(virtual)}}{p + \tilde{n}} \underset{\text{weak}}{\longrightarrow} \mu^+ + \nu_\mu$$

(see Fig. 4.19). Very rarely, a gamma ray accompanies the decay products. Naturally the above process is followed by the decay of the muon.

It is convenient to discuss at this point the fate of the neutral pion, π^0. It has a lifetime much shorter than its electrically charged siblings (a mere 0.8×10^{-16} s), suggesting a disintegration mechanism considerably stronger than the weak force: for example the electromagnetic force. This guess is confirmed by the nature of the decay products, which are usually two gamma rays:

$$\pi^0 \longrightarrow 2\gamma .$$

However, the π^0 is electrically neutral, so how can it decay via the electromagnetic force, to which it is apparently not even coupled? The resolution lies once again in virtual particles. The π^0 spends some of its time as a proton–antiproton pair, and these, as we know, can annihilate into two photons. We thus have the scheme

$$\pi^0 \underset{\text{strong}}{\longrightarrow} \underset{\text{(virtual)}}{p + \tilde{p}} \underset{\text{e.m.}}{\longrightarrow} 2\gamma .$$

One in every 86 decays produces a single γ together with an electron–positron pair. To understand this variant, envisage one of the photons as virtual, leading to a three-stage process (Fig. 4.20)

$$\pi^0 \underset{\text{strong}}{\longrightarrow} \underset{\text{(virtual)}}{p + \tilde{p}} \longrightarrow \underset{\text{(virtual)}}{\gamma + \gamma} \longrightarrow \gamma + e^+ + e^- .$$

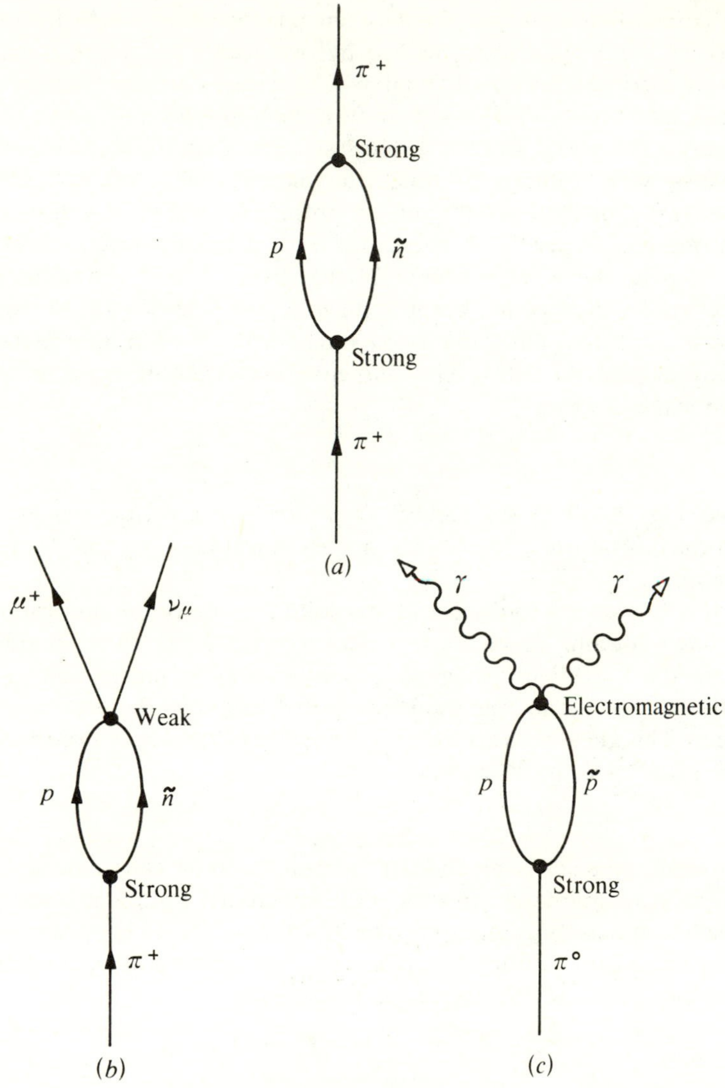

Fig. 4.19. Decay of the pion. For some of the time π^+ converts via the strong interaction into a virtual $p,\tilde{n}$ pair which then annihilates by either (*a*) the strong force (compare with Fig. 4.6) or (*b*) after many billions of attempts, the weak force. The upper vertex blob may be considered as a scaled-down version of a blob like that in Fig. 4.15. It conceals a *W* particle exchange. (*c*) A similar, though more rapid, fate for the neutral pion.

One in every 30 000 events produces *two* electron–positron pairs by this sort of scheme.

It is interesting to note that the first stage (strong interaction phase) of π^0 decay

$$\pi^0 \rightarrow \underset{\text{(virtual)}}{p + \tilde{p}}$$

is almost identical to the first stage of the charged pion decay

$$\pi^+ \rightarrow \underset{\text{(virtual)}}{p + \tilde{n}}.$$

This is because the strong nuclear force is the same between protons, neutrons and antiprotons. Thus the difference in the decay times of the charged and neutral pions can be attributed almost entirely to the *second* stage of the disintegration. It is because the electromagnetic force is several powers of 10 stronger than the weak force that the π^0 disintegrates more than 10^8 times more rapidly than the π^+ and π^-.

We have been rather cavalier in using the description 'the weak interaction' to cover a multitude of processes, which range in lifetime from 10^{-8} s for pion decay to 15 minutes for neutron decay. How can we be sure that we are dealing with the *same* weak force? Doesn't the existence of widely different lifetimes, and a plurality of neutrinos,

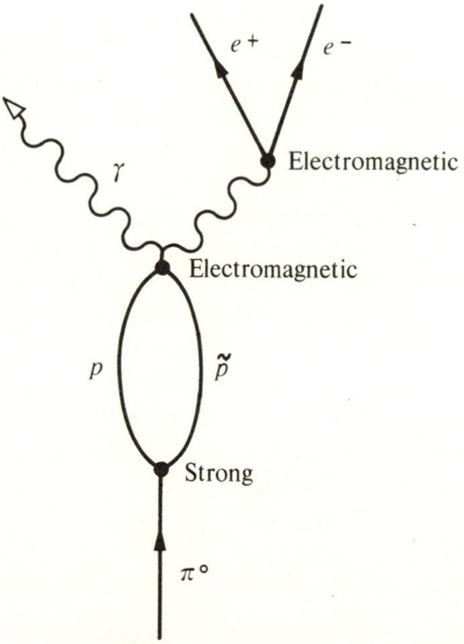

Fig. 4.20. Three-stage pion disintegration to a photon and an electron–positron pair.

suggest more than one weak interaction? Fortunately for the simplicity of nature this is not so. Detailed calculations using Fermi's model show that a number of quantities other than the strength of the weak force can affect the lifetime of these processes. For example, the energy available to drive the process, and the velocities of the final products, can drastically affect the speed of the reaction. Taking these effects into account satisfactorily explains the wide range of observed lifetimes.

There is also good evidence that the muon and its attendant neutrinos behave in all respects (except of course the muon mass) like the electron and its neutrinos; indeed the muon seems to be simply a big brother of the electron. Some of this evidence will now be described. Whenever an electron can enter into a reaction, so can a muon. Thus, for instance, we have the muon analogue of inverse beta decay, when a muon is captured by a proton

$$\mu^- + p \rightarrow n + \nu_\mu.$$

However, the spontaneous beta decay of the neutron into a muon, proton and neutrino ($n \rightarrow p + \mu^- + \bar{\nu}_\mu$) cannot occur through lack of energy.

The capture of a muon by a proton can occur in a curious way. Sometimes a free μ^- will encounter an atom and orbit around the nucleus for a while before it decays. This arrangement is referred to as a muonic atom. Because the muon is about 200 times heavier than an electron, it will orbit 200 times closer to the nucleus, which in a heavy atom is very close indeed, on account of the large nuclear charge. With this proximity, there is a good chance that the muon will (by the usual quantum uncertainty associated with position) find itself inside the nucleus for a moment, when it can interact with a proton and convert it to a neutron. On p. 102 we saw that electron capture also occurs by this mechanism. A careful comparison of the muon and electron capture rates confirms that the interaction strength is the same in both cases, which is good evidence that the muon and electron couple to other fermions through the same universal interaction.

A crucial test of the equivalence of the muon and electron is the decay of the pion. At first it seemed as though the pion always decays into a muon, but then it was discovered that one event in every 8000 or so produces an electron or positron

$$\pi^- \rightarrow e^- + \bar{\nu}_e$$

$$\pi^+ \rightarrow e^+ + \nu_e \,.$$

Very rarely, the decay products are accompanied by a gamma ray or even a π^0. The explanation of why the above decay mode is so rare is as follows. The disintegration energy of the pion is the same whether a muon or electron is produced. If it is an electron, its comparatively tiny rest mass means that a great deal of energy will be left over, which can appear as energy of the electron's motion. Thus the emerging electron is created moving very near to the speed of light. In contrast, the heavier muon uses up more energy to make its mass and is created moving relatively slowly. The importance of this is that the Fermi interaction depends sensitively on the speed of the decay products, and this operates to reduce the probability of the electron mode of decay compared to the muon mode.

Finally, a very elegant confirmation of the electron–muon special relationship comes from a study of their magnetic moments. Recall from Section 4.1 that the magnetic field of an electron (due to its spin) can be calculated exceedingly accurately, because it has no strongly interacting cloud of virtual charged pions around it to swamp the delicate electromagnetic correction. Only a cloud of virtual photons exists whose effect can be regarded, because of the relative weakness of the electromagnetic force, as just a small perturbation. Detailed computations can give the tiny corrections to the magnetic moments of the electron (see p. 130) and also the muon, to several significant figures, on the assumption that they differ only in their mass. Experiment has confirmed that these values are correct.

For a long time physicists have been perplexed as to the reason for the muon's existence. It is so like the electron that its role in nature would seem to be obscure. What purpose does it serve? In the next chapter we shall see how more recent theories of the weak and electromagnetic interactions perhaps provide a place for this enigmatic particle.

We end this section by briefly mentioning some of the experimental evidence for the existence of two separate types of neutrino. The first is some negative evidence concerning muon decay

$$\mu^- \rightarrow e^- + \nu + \bar{\nu} \ .$$

If the two neutrinos were each other's antiparticle then they could be expected to annihilate from time to time because they are created in close proximity to each other by the dying muon. But into what can a neutrino–antineutrino pair annihilate? One possibility is for them to form briefly a virtual electron–positron pair through the weak interaction, which then annihilates into a photon by the electromagnetic interaction. We thus have the three-stage process

$$\underset{\text{weak}}{\mu^- \rightarrow} e^- + \underset{\text{virtual}}{(\nu + \bar{\nu})}$$

$$\hookrightarrow_{\text{weak}} \underset{\text{virtual}}{e^+ + e^-} \underset{\text{e.m.}}{\rightarrow} \gamma$$

which amounts in the end to

$$\mu^- \rightarrow e^- + \gamma \, .$$

In spite of receiving much attention, this reaction has never been detected, which suggests that the middle link in the chain, $\nu + \bar{\nu} \rightarrow e^+ + e^-$, is forbidden. If the two neutrinos are distinct, this is explained.

An experiment to test the identity of the muon-neutrino was performed at Brookhaven in New York over a period of 25 days in 1962. A beam of pions was allowed to decay and pass through steel shielding into a special detector called a spark chamber. This is a device that registers the passage of a charged particle by causing the discharge of a spark, which can then be photographed. Electron-neutrinos when traversing ordinary matter are known to interact (though exceedingly weakly) with the neutrons and protons present in the nuclei, through the reactions

$$\nu_e + n \rightarrow p + e^-$$

and

$$\bar{\nu}_e + p \rightarrow n + e^+ \, .$$

Similarly, muon-neutrinos are expected to interact with n and p, but to yield muons in place of the electrons:

$$\nu_\mu + n \rightarrow p + \mu^-$$

$$\bar{\nu}_\mu + p \rightarrow n + \mu^+ \, .$$

If the two types of neutrinos are identical, then some of the muon-associated neutrinos produced from the pion decays, which happily produce muons via the second two processes, ought also to produce in the spark chamber *electrons* and *positrons* via the first two processes – at least on rare occasions. Such events were never observed.

Before leaving this section a few words should be said about the extremely ephemeral character of particles such as muons and pions that decay after periods of time that are minute by everyday standards. The π^0, for example, lives for only a ten-million-billionth of a second. It might seem surprising that we could know of the existence of such transient entities.

In subatomic physics, a good fundamental unit of time against which to gauge short durations is the time it takes for light to cross a proton or a neutron. This is a mere 10^{-24} s. On this scale, the lifetime of a muon, at 2×10^{-6} s, is almost infinite. During that time light can travel a few hundred metres. Frequently, particles are made which move nearly as fast as light, so a muon might well traverse the length of a large laboratory before decaying. This is more than long enough to experiment with it and study it in detail.

There is another factor which contributes to the relative longevity of unstable particles. The theory of relativity requires that the time scale of a particle which approaches the speed of light is *slowed down* relative to the laboratory. In macroscopic objects this time dilation effect enables a person who engages in near-luminal spacecraft travel to return to Earth many years younger than his twin left behind.* The dilation of time implies that although a muon has an average lifetime of only 2 μs in its own frame of reference, as observed in the laboratory, it can exist for much longer than this, and may travel many kilometres before disintegrating. Indeed, the first muons discovered were produced by cosmic ray bombardment far up in the atmosphere, yet travelled intact all the way to the surface of the Earth.

One of the most dramatic ways of observing the activity of subatomic particles is to use a *bubble chamber*, invented by the American physicist, Donald Arthur Glaser in 1952. This is a vessel filled with a transparent liquid that is superheated. This means it is in an unstable condition and will boil if disturbed, even microscopically. If an energetic electric particle ploughs through the liquid while in this condition it will ionize a trail of molecules and cause a string of tiny bubbles to form along its path. Glaser received a Nobel prize in 1960.

A precursor of the bubble chamber, called the cloud chamber, was invented in 1911 by the British physicist Charles Thomson Rees Wilson (1869–1959), who was awarded a Nobel prize for his invention in 1927. It operates by forming a trail of droplets in a supersaturated (damp) gas.

Let us summarize the developments of this chapter. We have discovered that there are at least four forces of nature. In order of increasing strength they are gravitation, weak, electromagnetic and strong. The first and third have long-range effects and were known in classical physics. The other two operate only over subatomic dimensions and require the new ideas of quantum field theory to make sense of them. These ideas describe a field of force in terms of quanta, or *particles*, which means that we cannot separate the study of the forces of nature from the study of elementary particles and the structure of matter. Even the internal structure of a seemingly isolated particle depends on a network of self-action through the cloud of virtual quanta which surrounds it. Thus, the forces alter the structure of the particles, and the particles communicate and mediate the forces.

* See Chapter 2 of *Space and time in the modern universe*.

The particle, or quantum, nature of force fields manifests itself conspicuously in the case of the strong, electromagnetic and weak interactions. The odd man out is gravity, whose strength is so feeble that there appears to be no hope whatever of directly observing quantum gravity effects (see Section 2.8). Nevertheless, one may conjecture that the gravitational field may be quantized in a fashion similar to the electromagnetic field, in which case the quanta of the field would be called gravitons. From general considerations it is known that the graviton would be a massless boson with a spin of two, as mentioned on p. 80.

Notice, however, that although the photon is the quantum responsible for mediating the electromagnetic force between electrically charged particles, it does not itself carry a charge. In contrast, the graviton does carry a 'gravitational charge' as well as acting as a communicator of gravitational force (similarly the pion is subject to the strong force). This means that 'gravity gravitates', which is to say that the graviton can be acted upon, and produce, a gravitational field of its own. This non-linear property has important consequences for the behaviour of strongly gravitating systems.

There seem to be six types of intrinsically stable particles in nature: electrons, protons, two types of neutrinos, the photon and the graviton. To this list we could also add the antiparticles of the first four (the photon and graviton are their own antiparticles). Antineutrinos are the only antiparticles that could be abundant in nature, at least in our galaxy.

Next we have the semi-stable neutron (and antineutron), and the intrinsically unstable mesons – the muons, the charged pions and the neutral pion which, like the photon, is its own antiparticle. The total, excluding antiparticles, is 10. (We also expect the existence of the *W* particles.) These 10 participate in some, but not necessarily all, of the four interactions; the table on p. 153 shows which, and summarize their properties.

There may have been a time in the mid-1930s when it seemed that a fundamental theory of matter was at hand. In such a theory all of matter would be built up out of aggregates of a few truly elementary particles. Physicists once had this hope for atoms, but the atom was shown to consist of nucleus and electrons. Then the nucleus itself was shown to be a composite body. For a while electrons and protons seemed adequate, but then neutrinos and neutrons came along. Finally the discovery of intrinsically unstable particles like the pion added to the general complexity. If the hope for simplicity still lingered until after the second world war, it was soon to receive a death blow.

Table 4.1. List of particles mentioned so far.

Particle	Symbol	Anti-particle	Strong	E.m.	Weak	Gravity	Mass	Spin	Lifetime
Proton	p	$\bar{p}$	Yes	Yes	Yes	Yes	938.3	1/2	∞
Electron	e^-	e^+	No	Yes	Yes	Yes	0.511	1/2	∞
Photon	γ	γ	No	Yes	No	Yes	0	1	∞
Electron-neutrino	ν_e	$\bar{\nu}_e$	No	No	Yes	Yes	0	1/2	∞
Muon-neutrino	ν_μ	$\bar{\nu}_\mu$	No	No	Yes	Yes	0	1/2	∞
Graviton	g	g	No	No	No	Yes	0	2	∞
Neutron	n	$\bar{n}$	Yes	No	Yes	Yes	939.6	1/2	918
Pion	π^+	π^-	Yes	Yes	No	Yes	139.6	0	2.6×10^{-8}
Pion	π^0	π^0	Yes	No	No	Yes	135.0	0	0.8×10^{-16}
Muon	μ^-	μ^+	No	Yes	Yes	Yes	105.7	1/2	2.2×10^{-6}

Mass in MeV, lifetime in seconds, spin in units of $\hbar$.

5

Law and order in the microworld

The discovery that pions can be liberated from nuclear particles under the impact of high-energy bombardment was a stimulus to build machines which could deliver fast-moving protons, electrons, pions and any other charged particles on to a selected target. These machines operate by accelerating charged particles to near the speed of light using electric and magnetic fields. In one type, called a cyclotron, the particles whirl round and round, building up more energy with each circuit. In another, the linear accelerator, pulses of electric field drive a beam down a long straight tube. In either case the principle is to deliver as much energy as possible to the particles and then smash them into something. The method is crude, but effective. Early machines were a few metres in size. Today, one of the world's largest cyclotrons (actually a close relative called the proton synchrotron) at Serpukhov in the Soviet Union is 472 m across and contains 20 700 tonnes of magnet, while the Stanford linear accelerator in California is over 3 km long. These technological dinosaurs are so powerful that they can accelerate particles to energies of many thousands of MeV. Reaching into the deepest recesses of matter, they burst open a whole new world of subatomic activity.

With the advent of the accelerators, it soon became clear that the muons and pions were not the only intrinsically unstable new particles, but merely the first of many. Year after year, more particles were discovered in profusion, until by 1955, 30 were known. Today the number stands at several hundred. How can we make sense of this bewildering array? Why should matter exist in so many varying forms? What relationship do all these particles have to one another and to the forces that they carry? Are there any truly *elementary* particles?

These are some of the questions that will be addressed in the next two chapters, though not necessarily answered. Our present situation is like that of the nineteenth-century chemists who, faced with a vast number of chemical compounds and dozens of elements, with diverse and confusing properties, slowly began to recognize patterns and regularities. Faith in the ultimate simplicity of nature eventually led to the organization of all the jumbled information in a systematic form. With Mendeleev's periodic table of the elements and the slow realization of atomic and molecular structure, order emerged from the chaos. Today we are back at the pre-Mendeleev stage as far as subatomic matter and forces are concerned. Patterns and regularities are apparent, but no real understanding has emerged. Ad hoc models abound. A good start has been made in the systematic organization of the data in a reasonably meaningful way, at least for some of the particles, but each year brings its own new crop of discoveries, new pieces for an incomplete jig-saw. Models are built and collapse with disturbing monotony.

Later in this chapter we shall examine some of the hopeful new ideas which have emerged in recent years, but first a survey of some of the new particles must be given, and then a glimpse at the way they are classified and ordered into families.

5.1 *Conservation laws and selection rules*

Scientists have long held a belief in the ultimate simplicity of nature. The world around us is full of rich and complicated activity and elaborate organization and structure, but by analysing the structure into progressively smaller units we expect to discover simple uniformity rather than complex variety. In the early days of atomic and subatomic physics this expectation may have appeared fulfilled. Electrons and protons are simple, uniform, and indistinguishable from their siblings. Furthermore they interact in an elegant, straightforward way through the operation of Maxwell's electromagnetism. Electrons and protons together build up all the complexity of matter that surrounds us.

This vision of an orderly microworld existing as a substratum beneath all material things did not last long. Indeed, it may appear to the reader that the organization of subatomic matter threatens to collapse into total chaos. Quantum theory undermines the substance of matter, allowing particles to emerge out of nowhere for fleeting moments only to disappear again. The concrete reality of an electron located at a place, carrying energy and momentum, is known to be a fiction; no one can know for certain where it is located or how it is moving. Instead of two or three basic elementary particles there are a seemingly endless number. Nor do they remain intact. Hundreds of different reactions occur that transmute them into one another. Far from simplicity, the diversity of the subatomic inhabitants seems as bewildering as the variety of life forms in a zoo. How can we make sense out of this muddle?

Fortunately, subatomic physics is not complete anarchy. Many conceivable transmutations and reactions simply do not occur at all. We do not see protons changing into positrons, or electrons into neutrinos. Muons do not decay into photons or neutrons into pions. Why not? What rules are there that bring at least partial discipline and order to the tangle?

Classical physics has rules governing the interactions of matter. For particles, the most obvious of these are the laws of energy and momentum conservation which tell us that the total energy and total momentum must be the same before and after any change (See section 1.1). When relativity is taken into account, we must include mass with energy because of the equivalence $E = mc^2$. Less familiar is the conservation of angular momentum or spin. As we frequently deal with spinning subatomic particles this rule is very important. As explained in Section 2.7, spin angular momentum and that due to the motion of a particle are intrinsically different, so we expect them both to be conserved, which is indeed found to be so.

The conservation laws of mass-energy, momentum and spin are all examples of *dynamical* conservation laws, because they have to do with the motions of material bodies. There are also laws which require the conservation of other quantities, unconnected with motion, such as electric charge. No experiment has ever succeeded in creating or destroying electric charge (of course, it is understood that positive charge can be created along with an equal quantity of negative charge). In subatomic physics, charge always comes in multiples of the basic unit, and usually in only the lowest multiple – the charge on the electron. The law of charge conservation tells us that the total number of units of positive electricity must be the same before and after the transmutations.

Let us see how the processes discussed so far obey these various rules of conservation. The first transmutation, mentioned in Chapter 3, was the decay of the neutron

$$n \rightarrow p + e^- + \tilde{\nu}_e .$$

First, is there enough mass-energy available? The neutron has mass 939.57 MeV, the proton 938.28 MeV and the electron 0.51 MeV. The antineutrino is massless. Thus the right-hand side adds up to 938.79 MeV, so the neutron has plenty of mass-energy to spare – more than an electron mass – which, according to the laws of quantum mechanics, is divided randomly among the three product particles. Some is carried away by the neutrino, the rest appears as the energy of motion of the proton and electron. In beta decay the electron may be ejected at speeds close to that of light.

Neutron decay also conserves spin: all four particles have spin one half, so by creating two of the product particles with their spins directed opposite to each other to cancel, the remaining particle can conserve the spin one half of the neutron. Electric charge conservation is similarly respected. The neutron and neutrino are chargeless, and the single unit of positive charge on the proton cancels the unit of negative charge on the electron.

Interchanging incoming particles for outgoing antiparticles always respects the charge and spin conservation laws, but may violate energy conservation. Thus from the reaction displayed above we obtain

$$p + e^- \rightarrow n + \nu_e$$

but the combined rest mass on the left is less than the neutron rest mass, so this reaction is an 'uphill' process. Consequently it will not occur if the proton and electron are at rest. However, if they are collided at speed, when their energy of motion can make up the deficit, or when the proton is bound in a nucleus so that the binding energy can contribute, then the reaction will proceed (it is, in fact, a mode of beta decay – see p. 102).

As another example, consider the decay of the charged pion

$$\pi^- \rightarrow \mu^- + \bar{\nu}_\mu \rightarrow e^- + \bar{\nu}_e + \bar{\nu}_\mu + \nu_\mu.$$

The masses in MeV of π^-, μ^- and e^- are 139.57, 105.66 and 0.51, respectively, so there is an overabundance of available energy. The pion has spin zero, the others all spin one half. The μ^-, $\bar{\nu}_\mu$ pair are created spinning in opposite directions, so cancel their total spin to zero, as required. Then the muon decays to e^-, $\bar{\nu}_e$ and ν_μ, and the final four particles cancel all their spins pairwise. There is one unit of negative charge throughout.

In all these examples, momentum is also conserved. A good example of momentum conservation has already been described on p. 109. This law forbids the annihilation of a particle–antiparticle pair into a single *photon*, e.g. $e^- + e^+ \nrightarrow \gamma$.

In addition to the four classical laws that describe how various quantities remain unchanged during all natural activity, there is also a universal *asymmetrical* law which regulates the *organization* of the activity. This law is just as important as the conservation laws for controlling events in the microworld. In the macroscopic world the organization of matter and energy is subject to the so-called second law of thermodynamics. This law can be stated in many different ways and has universal application to systems as diverse as stars, gases, crystals and human beings. Stated roughly, the second law requires that in any change, the universe becomes a little more disordered. Thus, in contrast with the constancy requirement of a conservation law, this law says that disorder must always become greater and greater whenever any natural process occurs.*

Very often it is possible to express the increase of disorder in terms of energy. A system which contains a lot of ordered energy has a natural tendency to disintegrate into a chaotic state in which the energy gets spread about randomly, and dissipated. Good examples are the human body and TNT. In both cases a lot of energy is ordered in one place; unless carefully protected, they are likely to disintegrate more or less violently, into disordered and useless energy.

In microscopic physics the same principle regulates the arrangement of energy among subatomic particles. The decay of an excited atom or a radioactive nucleus are examples of how ordered energy, locked up inside the atom, dissipates away into the universe in the form of photons, or alpha and beta rays.

The tendency for energy to 'break out' and escape away into the surrounding environment is always evident. One of the most conspicuous examples of energy being locked up is *mass*. The mass of a particle represents highly ordered energy, and given half a chance it will disintegrate violently. This is what happens when an electron and positron meet; the energy confined in the particles' mass is liberated and flies away in the form of gamma rays. It happens also in the decay of the neutron, the pion or the muon. Always the energy tries to erupt from the imprisonment of a massive particle and spread itself among less massive particles. Sometimes there is a whole sequence of disordering transitions. Consider, for example, the annihilation of a proton–antiproton pair into pions, followed by pion and muon decays:

1. Pions produced $p + \bar{p} \rightarrow \pi^+ + \pi^- + \pi^0$
2. Pions decay $\mu^+ + \nu \quad \mu^- + \bar{\nu} \quad 2\gamma$
3. Muons decay $e^+ + \nu + \bar{\nu} \quad e^- + \nu + \bar{\nu}$
4. Positron annihilates with random electron $e^- + e^+ \rightarrow 2\gamma$.

What began as 1876.6 MeV energy locked up quietly in the masses of p and $\bar{p}$ has ended up spread among no less than 11 particles, 10 of which shoot off in different directions at the speed of light. The remaining electron careers around at high speed, dissipating its excess energy in the form of yet more photons. The end result has been to convert ordered mass-energy into disordered radiation; a tiny residue of the original mass-energy survives in the single electron.

In all cases, the natural tendency is for a massive particle to disintegrate into several less massive ones. The cascade stops when no mass is left. For a particle to possess mass means that it is poised on the

* For a detailed exposition see *Space and time in the modern universe.*

edge of disaster. How, then, is it possible for electrons and protons to retain their mass without suffering explosive decay? First, because the universe – at least our galaxy – seems to be composed of matter rather than antimatter. In the absence of positrons or antiprotons the electrons and protons are stable. Why?

The stability of the electron is easily explained. The only spin one half particle that is lighter is the neutrino, which is electrically neutral. Thus, charge conservation saves the electron. Neutrinos and electrons sit at the bottom of two separate ladders of disintegration, one for neutrals, one for electric particles.

Coming to the proton the situation is less clear. Why can it not decay into a positron and, say, a photon:

$$p \rightarrow e^{+} + \gamma .$$

This reaction preserves all the classical conservation laws: energy, charge, spin and momentum. It is also in accordance with the natural tendency towards disorder. The proton is perched precariously high above the positron, with an enormously greater mass trapped inside it. What prevents this mass-energy from breaking out?

Unless there exists some rule of nature which forbids proton decay, then we know from the statistical properties of quantum mechanics that decay is bound to occur with a definite probability. It seems that everything in the microworld which nature allows, actually happens. As the classical conservation laws are unable to stop the decay of the proton, there must be a new type of conservation law whose operation is unknown in the everyday, macroscopic world, but which exercises powerful control over submicroscopic, quantum processes. Suppose the proton possesses some new kind of quality, analogous to electric charge, which is absent in the positron. We could then explain why the proton does not decay into the positron: there is a law of nature that says the new quality must be conserved. The name invented for this analogue of electric charge is *baryon number* (baryon means 'heavy one'). Like charge, baryon number comes in whole units, and the proton has +1 of them. Also like charge, the antiproton has baryon number −1. All lighter particles have baryon number zero: they are baryon-neutral. Particles that carry non-zero baryon number are known collectively as *baryons*.

When a proton and antiproton come together, their combined baryon number is zero, so the law of baryon conservation no longer prevents decay into lighter, baryon-neutral, particles:

$$p + \bar{p} \rightarrow \pi^{+} + \pi^{-} + \pi^{0}$$

or $$\rightarrow 2\gamma$$

or $$\rightarrow \nu + \bar{\nu}, \text{ etc.}$$

In the transmutation of the neutron into a proton

$$n \rightarrow p + e^- + \bar{\nu}_e$$

(or vice versa) the right-hand side has baryon number +1, so conservation demands that the neutron too carries +1 units. The antineutron has −1 units.

Baryon number differs from electric charge inasmuch as it does not seem to be associated with a long-range field of force. We cannot detect baryon field waves or baryon forces like electromagnetic waves and forces. Although it may seem somewhat contrived, the law of baryon conservation is of vital significance. If it were ever to fail, the entire material contents of the universe would disappear in a fireball of gamma radiation, as the protons decayed to positrons and annihilated all the electrons – an awesome prospect in the extreme. The experimental evidence for baryon conservation is excellent. Even if the proton were only minutely unstable, there are so many of them around in ordinary matter that the occasional decay of one of them would be detectable. Measurements show that their half-life must be at least 2×10^{30} years – a hundred billion billion times longer than the present age of the universe – for their decay to go unnoticed.

Powerful though it is, baryon conservation cannot explain the organization and activity of the lighter particles which are baryon-neutral. Consider, for example, neutrinos. They are the simplest of all particles, having no mass, charge or strong interaction – only spin. How can there be *four* different types of such a faceless entity? What features can there be to distinguish an electron–neutrino from a muon-neutrino, or either of these from their antiparticles? Yet nature manages to tell them apart. If a neutrino strikes a neutron, there is an exceedingly tiny probability that it will transform it into a proton through the weak interaction process

$$n + \nu \rightarrow p + e^-.$$

This conversion does not seem to occur, however, using antineutrinos

$$n + \bar{\nu} \nrightarrow p + e^-$$

so evidently the neutron can tell the difference.

This fact can be explained if the neutrino carries some sort of label, similar to baryon number, which reveals its identity to the target neutron. The name given to this second mysterious microscopic quality is *lepton number*. (Lepton means 'light thing'.) Like baryon number and electric charge, lepton number comes in whole units: neutrinos have +1 unit and antineutrinos −1. Electrons and muons also possess lepton number: e^- and μ^- have +1 unit, e^+ and μ^+ have −1 unit. The particles which carry non-zero lepton number are known collectively as *leptons*. They are all the lightest spin one half particles (muons, electrons, positrons and neutrinos of all types).

By inverting another subatomic conservation law – the conservation of lepton number – reactions such as $n + \tilde{\nu} \rightarrow p + e^-$ are automatically forbidden because the $\tilde{\nu}$ on the left-hand side has lepton number -1 whereas e^- on the right has $+1$. On the other hand the neutron recognizes the neutrino with the correct -1 lepton units and allows the reaction $n + \nu \rightarrow p + e^-$ to proceed. Thus, lepton number is an important quality to add to the rather sparse list of properties possessed by the ghostly neutrinos.

It is interesting that lepton number conservation provides a curious argument for the exact masslessness of the neutrino. The reason for this concerns the nature of the spin of neutrinos. According to quantum theory (see p. 75) if a particle has a spin of one half then the spin axis can only point in one of two directions. The direction of motion of the neutrino defines a special direction, so we can envisage the neutrino as travelling with a sort of corkscrew motion (see Fig. 5.1). Experiment shows that it is a left-handed screw. The anti-neutrino is only distinguished from it dynamically by the fact that it spins the opposite way (right-handed screw). If these particles had a mass, however small, then they would be obliged by the theory of relativity to travel slower than the speed of light. Consequently, it would be possible to chase after them and overtake. As we passed by such a neutrino it would be moving, relative to us, in a backward direction. So instead of being a left-handed screw it would now be right-handed: in short, an antineutrino. Thus, what to one observer appears as a neutrino, to another

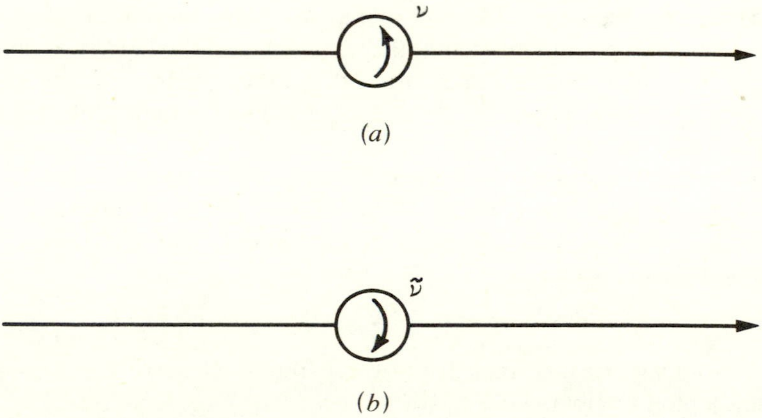

Fig. 5.1 Neutrino corkscrew. (*a*) The neutrino always spins like a left-handed screw. (*b*) The antineutrino spins like the mirror image of the neutrino.

observer moving differently appears as an antineutrino. We could convert one into the other simply by overtaking it. According to relativity, physics must be independent of relative motion, so these two particles would have to be indistinguishable in their reactions, which we know would be incorrect as it violates lepton number conservation. However, if the neutrino is massless, it will travel at the speed of light, and no observer in the universe can move fast enough to overtake it. Only massless neutrinos preserve their spin identity in all reference frames.

An important application of lepton number conservation is to the decay of the muon

$$\mu^- \rightarrow e^- + \tilde{\nu}_e + \nu_\mu .$$

The left-hand side has lepton number +1 units, and the right has +1 −1 +1 = +1 units, as required for conservation. Another mode of muon decay permitted by all the conservation laws would be

$$\mu^- \rightarrow e^- + \gamma$$

in which a photon is produced instead of the neutrinos. This process was discussed on p. 150, where it was pointed out that extensive searches have failed to produce any evidence of the gamma decay of the muon. It is therefore supposed that it must be forbidden. We can explain this by still another conservation law: conservation of *muon* lepton number. The idea is that the electron-neutrino (ν_e) carries a *different type* of lepton number from the muon-neutrino (ν_μ), and that each kind of lepton number is *separately* conserved. In this scheme μ^- has a muon lepton number of +1 (μ^+ has −1) and zero electron lepton number, whereas e^- is vice versa. The gamma decay of the muon is therefore forbidden because it would mean a change of muon lepton number +1 into electron lepton number +1, thereby violating the laws of separate lepton number conservation. On the other hand, the reaction $\mu^- \rightarrow e^- + \tilde{\nu}_e + \nu_\mu$ preserves a total of +1 and 0 muon and electron lepton units, respectively.

If the reaction $\mu^- \rightarrow e^- + \gamma$ were the only one to be forbidden under the separate laws, the concept of two distinct types of neutrinos and antineutrinos would be vacuous. It turns out, though, that other reactions, such as

$$n + \nu_\mu \rightarrow p + e^-$$

are forbidden because they do not separately conserve electron and muon lepton numbers (see p. 150).

It is interesting to note that all the fermions mentioned so far carry either baryon or lepton number, whereas the bosons are both baryon-

and lepton-neutral. One consequence of this is that leptons and baryons can only be created or annihilated in *pairs*, in order to conserve total lepton and baryon number. In contrast, photons and pions can come and go in any number.

To sum up the results of this section, we can say that all subatomic particles carry a number of labels which sort them into different species and are closely associated with the way in which they behave and interact. Some of the labels, such as mass, charge and spin, are just scaled-down versions of qualities familiar in the macroscopic world (though the 'double-turn' nature of intrinsic spin has no exact counterpart). On the other hand, qualities like baryon and lepton number are mysterious to us because they have no everyday manifestation. It is as though we cannot read these labels with our macroscopic apparatus; they do not produce any forces or fields which we can detect. Nevertheless, subatomic particles can read the labels perfectly, and respond accordingly.

The labels help particles to recognize each other and know how to react. All particles interact with all others to a greater or lesser extent, varying from the softest brush of the neutrino to the explosive annihilation of a proton with an antiproton. Label reading by particles is like a system of senses. Some of them (e.g. neutrinos) are blind to electric charge but see lepton number. Others (e.g. proton) see electric charge and baryon number. The photon sees only electric charge.

The conservation laws bring partial order to the chaos. There are still a great many ways in which the different particles may react together, but there is a hint of organization about it. Many processes are forbidden and others are regulated using the labels. All four forces of nature seem to respect these conservation laws absolutely. If the various particles discussed so far were the only ones in existence, perhaps we should be content to accept these laws as a complete theory of matter and the forces of nature, and probe no further. However, the bewildering discoveries of more and more species of particles not only calls for other new labels, it greatly increases the range and complexity of subatomic activity.

5.2 *Symmetry*

Symmetry has long played an important role in art, religion and mathematics. In nature too symmetry abounds, though in many physical systems it is hidden by their complexity. Throughout the development of physical theory the fundamental significance of symmetry has been recognized, but in recent years its power and elegance has led to some amazing advances in our understanding of the organization of subatomic matter. As a unifying principle in physics it is without equal.

A proper analysis of symmetry involves advanced mathematics and is beyond the scope of this book, but the basic ideas are familiar enough and easy to grasp. To fix ideas, consider some simple geometrical line shapes: square, equilateral triangle, circle (see Fig. 5.2). Each has rich and interesting symmetry properties. Perhaps the most familiar, possessed by all three, is *reflection symmetry*. If a mirror is placed perpendicular to the page along the broken lines (try this), the shapes remain unchanged. In each case, the left-hand side of the figure is the reflection of the right-hand side. A fancy way of expressing this is to say that the figures remain *invariant* under reflections in the broken line axes. Notice that several axes of reflection symmetry exist for each figure: four for the square, three for the triangle and an infinite number for the circle (placing the mirror along any diameter will do).

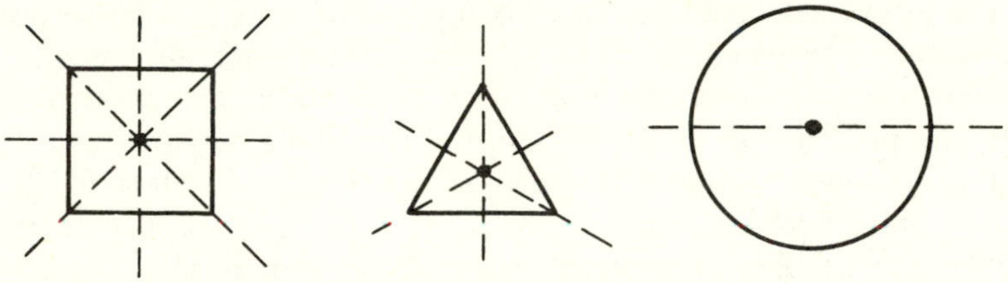

(*a*) (*b*) (*c*)

Fig. 5.2. Examples of geometrical symmetries. Each figure shown is unchanged if reflected in one of the broken lines.

We can find other symmetries in these figures. If the triangle is rotated through 120°, 240° and 360° about the dot in the middle, it looks identical. The square can be rotated to four positions 90°, 180°, 270°, 360° and remain the same. We say that they are *invariant* under rotations through multiples of 120° and 90°, respectively. Coming to the circle, it is clear that its symmetry is a whole order more powerful. *Any* rotation whatever about its centre leaves the circle invariant; indeed, the circle *is* the path of a rotating point. This comparison illuminates the first basic subtlety concerning symmetry. It comes in two rather distinct varieties: continuous and discrete. The rotation of the circle is a continuous operation, at all times leaving the shape unchanged. On the other hand, the rotations of the square and triangle, or the reflection symmetries, are discrete because they are isolated jumps, between which the symmetry is not maintained.

There is a second basic subtlety. In the case of the rotations, the same effect could be achieved by leaving the figures alone and *rotating the observer*; you can either turn the book beneath your eyes, or leave the book and walk around it. In contrast, the reflection operation cannot be carried out by moving the observer. We therefore see that there can be two types of symmetry operation: those that can be physically implemented and those that cannot. (Using a mirror is a cheat, of course. We are not looking at the *same* figure when the mirror is in place. Its use is simply an aid to visualization.)

An interesting perspective on symmetry is obtained if we ask what it is that really characterizes the superior symmetry of the circle over that of, say, the square. One way of looking at this is to see that the square has more structure than the circle. Compared to a square, a circle is rather featureless. We could destroy the rotation symmetry of the circle by flattening it a bit, or painting a dot on it. In both cases the result is to add new features and structure. Systems with very few features usually possess powerful symmetries.

The classic example of a featureless system is empty space. At first sight it appears to have no features at all, but that is not correct because it certainly possesses properties such as distance and angle. Nevertheless it seems that these properties remain invariant under, for example, continuous rotations. To see this, imagine that instead of rotating space (which is impossible) the observer rotates. In all directions the featureless void looks the same.

Empty space also possesses another continuous symmetry. Not only does it look the same in all directions, it looks the same at all places. One way of expressing this is to say that if space (of the observer) is *translated* (i.e. moved sideways) by any amount, everything is unchanged. In short, empty space is invariant under continuous rotations and translations. It is also invariant under reflections, of course. Empty space in a mirror is no different from empty space.

In this featureless, empty world, time also possesses symmetry. In a void in which nothing happens, one moment of time is as good as any other. There is nothing to distinguish them, so there is invariance under time translations, too. There is also invariance under temporal reflections, i.e. *time reversal*. As there is no activity, there can be no distinction between past and future time directions.

In the real world, space is not totally empty. There are fields and particles and complex activity. The symmetries which are exactly true for empty space are only approximately true, or not at all. For example, in the solar system all directions are not equivalent, because in the direction of the sun things look very different from the opposite direction. However, for many purposes the disturbances to exact symmetry presented by these features are unimportant and can be ignored to good approximation.

To be specific, consider an isolated particle residing somewhere in outer space. The particle could be either a billiard ball or an atom (quantum effects will be ignored), but we shall suppose that bodies such as the sun, as well as other particles, are too far away to have much influence on the behaviour of the particle, and effects of any fields are negligible. If the particle were suddenly to fly off in a particular direction we should be surprised and suppose that some external force had been overlooked. We are confident that in the absence of all forces the particle would not move. The basis for this confidence is precisely our unquestioning assumption that space is *symmetric* under translations. If one part of space is the same as every other, why should one place be distinguished by the sudden arrival of a particle? Moreover, why should the particle choose one particular direction in which to fly off, rather than any other?

Similar reasoning can be applied to rotations. We should not expect a body suddenly to start spinning without external propulsion, for why should it spin, say, clockwise rather than anti-clockwise? In addition, a body spins about an axis, which defines a special direction in space. If space is symmetric under rotations, no direction is special. Hence we do not expect a body to start spinning spontaneously.

These crude observations can be made mathematically precise and provide a deep and powerful connection between, on the one hand, the geometrical symmetries of space, and on the other, the *dynamical behaviour* of material bodies. Specifically, forbidding the absence of sudden spontaneous changes in motion amounts to a statement of the *laws of conservation of momentum and angular momentum*. The translation symmetry of space leads directly to momentum conservation for particles, whereas the rotational symmetry implies angular momentum conservation. In addition to this, the conservation of *energy* can be shown to follow from the translation symmetry of time (one moment is as good as any other). Thus, the most fundamental and comprehensive laws of physics are seen to follow from the elementary and unsurprising fact that empty space and time are featureless. What better example could one want to demonstrate the staggering power of symmetry in ordering the natural world?

These symmetries of space and time have been built almost unconsciously into the laws of physics. Newton's mechanics, quantum mechanics and relativity all possess the characteristic that their laws of motion are *invariant* under the symmetry operations such as rotations and translations. In the same way, physicists never really doubted that the laws of physics should incorporate the other symmetries we have discussed – the discrete symmetries, such as space reflection and time reversal.

As mentioned above, the status of these symmetries is a little different from that of the continuous symmetries, because we cannot actually implement them. We cannot produce a 'mirror universe' or a 'time-reversed' universe. Of course, we cannot rotate or translate space either, but we can rotate and translate the *observer* and that is usually equivalent as far as symmetry is concerned. It is not possible, though, to construct a 'mirror observer' or one whose time is reversed relative to us.

Although space and time reflections cannot actually be implemented the laws of physics can still be tested to see whether they remain invariant under these hypothetical operations. The way to approach this is to ask: suppose a movie film is taken of some particular natural activity and then projected into a mirror (or with the film back to front), would we notice the deception? Is the mirror image obviously an impossible process? Likewise, if the film is run backwards, will the process appear to violate the laws of physics?

To take a simple example, suppose the film shows a spinning sphere (see Fig. 5.3). The spin axis defines a particular direction and we can draw a line through it. If the spinning sphere is seen in a mirror its 'handedness' is reversed – clockwise and anti-clockwise become interchanged. We see a sphere spinning the opposite way. Clearly there is nothing objectionable about a sphere spinning the opposite way and we should not denounce the mirror image as a physical impossibility. Of course, if on closer inspection the sphere was seen to be the Earth, the deception would be exposed because we should see the dawn coming across the continents from west to east instead of east to west. But what we are supposed to be testing is the symmetry of the *laws* of physics, not the symmetry of the real world. In the world of subatomic particles, there are no 'continents' to distinguish one particle from another, so these incidental complications do not arise. In Section 5.4 some experiments will be described in which subatomic processes have been checked for mirror symmetry.

It is interesting to note that the existence of isolated magnetic charges could lead to a violation of space reflection symmetry. To see this consider the magnetic field around a current-carrying wire. A small, pivoted magnet will line up N–S, perpendicular to the wire, as shown in Fig. 5.4*a*. In a mirror the N–S direction seems reversed and we recognize an impossible situation: the line-up of the magnet could never look like this in the real world, because it would mean the north pole of the magnet was attracted rather than repelled by the north pole of the field. However, on closer inspection we see that the magnetism in the small bar is produced by tiny circulating currents, and these also reverse direction in the mirror – clockwise currents change to anti-

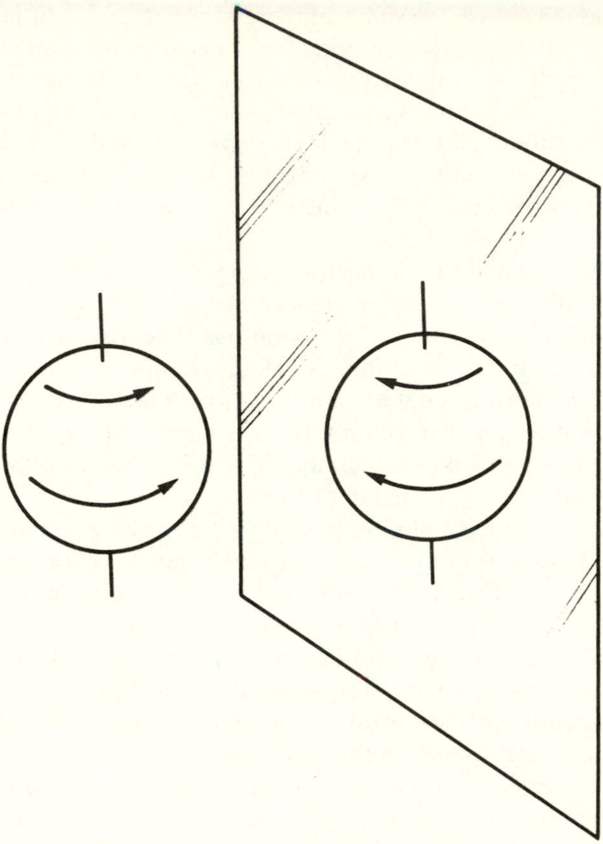

Fig. 5.3. Reflection symmetry. The spinning sphere rotates clockwise in the real world and anti-clockwise in the mirror world. The latter is unexceptionable, and if we didn't see the edges of the mirror we could not tell which sphere was real and which was an image. Both are equally possible situations. The mirror view would also appear if a movie film of the spinning sphere was played backwards.

clockwise currents. As a result the north and south poles become interchanged, and everything looks all right after all. Evidently electricity alone is incapable of distinguishing left from right.

Suppose now that instead of creating the magnetism from circulating currents, we made the magnet from two free magnetic poles, N and S, stuck to opposite ends of a bar (see Fig. 5.4*b*). The mirror image would now appear quite wrong, with the roles of the N and S poles reversed. Similarly, a *current* of magnetic poles would cause electric charges to move in the 'wrong' direction in the mirror image. So if electricity and 'magneticity' could be simultaneously present, left- and right-handedness would be clearly distinguished.

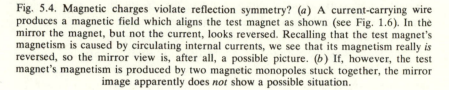

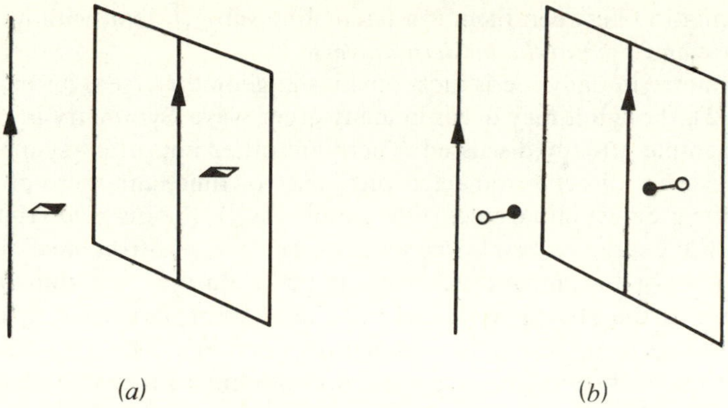

Fig. 5.4. Magnetic charges violate reflection symmetry? (*a*) A current-carrying wire produces a magnetic field which aligns the test magnet as shown (see Fig. 1.6). In the mirror the magnet, but not the current, looks reversed. Recalling that the test magnet's magnetism is caused by circulating internal currents, we see that its magnetism really *is* reversed, so the mirror view is, after all, a possible picture. (*b*) If, however, the test magnet's magnetism is produced by two magnetic monopoles stuck together, the mirror image apparently does *not* show a possible situation.

The example of the rotating sphere also serves to illustrate time-reversal symmetry, for a movie film played backwards will also show a reversal of the spin. Unless we know about the internal workings of the movie equipment we cannot tell which film sequence is forwards and which backwards – both look equally possible situations if the sphere is relatively featureless. Although in the everyday world of more complicated systems we usually can spot immediately the deception if a film is run backwards, in the microworld there appears nothing remarkable in the reversal of spin. The same applies to other familiar processes, such as the collision and disintegration of particles – their time reversals do not appear miraculous either. Only when the activities of many particles together are reversed do we suspect something. For example, the spontaneous disintegration of a neutron into p, e^- and $\bar{\nu}_e$ if shown in reverse would invite scepticism because it would show the exceedingly improbable triple encounter of a proton, electron and neutrino. For macroscopic processes, the odds against their reversal occurring are astronomical.

There is no contradiction between the time-reversal invariance of the laws of physics and the time *asymmetric* tendency, mentioned in the previous section, for all systems to dissipate their energy and seek the most disorderly arrangement. The former refers to the laws governing the *individual* particles, the latter to their *collective organization*. The reconciliation between them is a fascinating subject dealt with in detail in *Space and time in the modern universe.*

Symmetry in daily life is most obvious in geometry (see, for example Fig. 5.2), though it may occur in many other ways. Symmetry in time is one example already discussed. There are other important symmetries in physics not directly connected with space or time, and these prove to be of the greatest importance. One simple case is the discrete symmetry of electric charge reversal. The electron and the positron have already been described as 'mirror' particles, and in a sense we can think of the positron as the charge 'reflection' of the electron (we have to reflect other things too, like spin and lepton number). The fact that the magnitude of the charge is the same for both means that they are exact reflections of one another: a simple reversal of the *sign* of the charge is all that is necessary. Just as with space and time reflections, so we might expect the laws of physics to be invariant under charge reflection too. In later sections we shall see how each of these three symmetries holds up in practice. However, there is an interesting mathematical theorem which proves that, subject only to some very weak assumptions which nobody seriously doubts, the laws of physics *must* be invariant under the *combined* operations of space reflection, time reversal and charge reflection. The operations are usually denoted by P, T and C, respectively (P is used for space reflection because this operation is also known as 'parity' change). The theorem is known as the PCT theorem.

The PCT theorem provides the justification for regarding the emission of a particle as equivalent to the absorption of an antiparticle, a fact which we have used in previous chapters. The operation C changes particle to antiparticle, T changes emission to absorption, and P changes a particle which arrives from the right into one which arrives from the left. Fig. 5.5 shows an example of the PCT theorem at work.

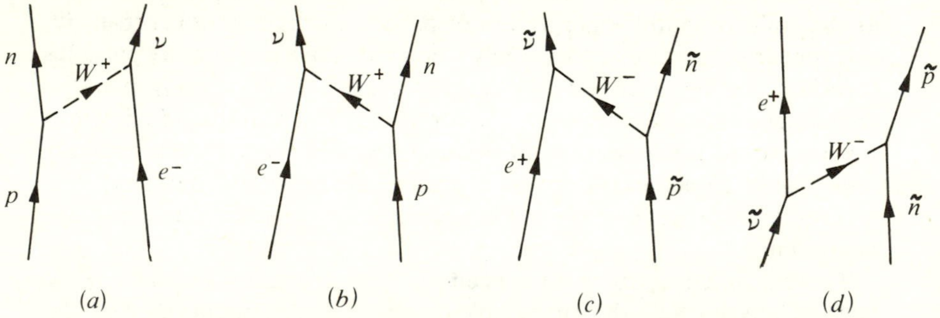

Fig. 5.5. The PCT theorem at work. Process (*d*) may be obtained from (*a*) by the successive operations of mirror reflection P, shown in (*b*), charge reversal C (particle → antiparticle) shown in (*c*) and time reversal T. The PCT theorem guarantees that if (*a*) is an allowed process, so is (*d*).

5.3 *The strange particles*

In the late 1940s the activity of unknown particles was observed from cosmic ray showers. In 1947 George Rochester and Clifford Butler of the University of Manchester obtained cloud chamber photographs showing V-shaped tracks made by the decay products of some new, heavy particles. Typically, the new particles are produced by smashing negatively charged pions into protons

$$\pi^- + p \rightarrow \text{ something new.}$$

The result, as seen in the cloud chamber (see Fig. 5.6), is the abrupt disappearance of the pion, a gap of a few centimetres where nothing at all is detected, and then some sets of V-shaped tracks caused by protons and pions. This sequence of events is easily interpreted as the production of unknown, electrically neutral particles which subsequently decay into pions and protons. The electrical neutrality of the intermediate objects explains the absence of a cloud chamber track.

We now know that there are actually *six* different kinds of these new particles, and unravelling all their properties was a major preoccupation of particle physicists in the 1950s. Two of the particle types are mesons – heavier than electrons but with zero baryon number. These new mesons are big brothers to the pion. First there are the kaons, or K mesons, that can either be charged or electrically neutral. Then, on its own, is the still heavier eta meson (written η) which is electrically neutral.

Perhaps more astounding than the new mesons is the existence of four sets of particles which are *heavier* than even protons or neutrons. In ascending order of mass they are lambda (Λ) which is neutral, sigma (Σ) with both charged and neutral members, xi (Ξ), once called the cascade particle, which also comes in charged or neutral varieties, and finally the charged omega (Ω). Collectively these heavy particles are sometimes referred to as *hyperons*. The new particles are shown together with the old in a table on pp. 172–3. Their discovery greatly widened and deepened the search for an understanding of the complexities of fundamental matter.

All the new particles participate in the strong nuclear force, and the proliferation of such entities demands a collective name. The name chosen is *hadron*. A hadron interacts strongly with other hadrons, and will also interact with them through a variety of other forces. The strong force is often called the *hadronic interaction*. Hadrons can be both bosons and fermions. The *mesons* are always bosons; pions, kaons and η all have spin zero. The hyperons (as well as the two nucleons, n and p) are all fermions. Λ, Σ and Ξ all have spin one half. The Ω is a surprise: it is a fermion with spin *three* halves.

Table of long-lived particles*

Family groups	Particle	Symbol	Mass	Charge states	Spin	Strange-ness	life-time	Principle decay products
	Graviton	g	0	0	2	0	∞	Stable
	Photon	γ	0	0	1	0	∞	Stable
Leptons	Neutrinos	$\nu_e\ \bar{\nu}_e$	0	0	1/2	0	∞	Stable
		$\nu_\mu\ \bar{\nu}_\mu$	0	0	1/2	0	∞	Stable
	Electron	$e^+\ e^-$	0.511	+1 −1	1/2	0	∞	Stable
	Muon	$\mu^+\ \mu^-$	105.66	+1 −1	1/2	0	2.2×10^{-6}	$e\nu\bar{\nu}$
Mesons	Pion	$\pi^+\ \pi^-$	139.57	+1 −1	0	0	2.6×10^{-8}	$\mu\nu$
		π^0	134.96	0	0	0	0.8×10^{-16}	2γ
	Kaon	$K^+\ K^-$	493.67	+1 −1	0	+1 −1	1.2×10^{-8}	$\mu\nu,\ \pi\pi^0,\ \pi^\pm\pi^+\pi^-$
		$K^0\ \bar{K}^0$	497.67	0	0	+1 −1	0.9×10^{-10}	$2\pi^0,\ 3\pi^0,\ \pi^+\pi^-$

Hadrons			Eta	η	548.8	0	0	0	2.5×10^{-19}	2γ, $3\pi^0$, $\pi^0\pi^+\pi^-$
	Baryons	Nuc-leons	Proton	$p\ \bar{p}$	938.28	+1 −1	1/2	0	∞	Stable
			Neutron	$n\ \bar{n}$	939.57	0	1/2	0	918	$pe\nu$
		Hyperons	Lambda	$\Lambda\ \bar{\Lambda}$	1115.60	0	1/2	−1 +1	2.6×10^{-10}	$p\pi$, $n\pi^0$
			Sigma	$\Sigma^+\ \bar{\Sigma}^+$	1189.37	+1 −1	1/2	−1 +1	0.8×10^{-10}	$p\pi^0$, $n\pi^{\pm}$
				$\Sigma^0\ \bar{\Sigma}^0$	1192.47	0	1/2	−1 +1	5.8×10^{-20}	$\Lambda\gamma$
				$\Sigma^-\ \bar{\Sigma}^-$	1197.35	−1 +1	1/2	−1 +1	1.5×10^{-10}	$n\pi^{\mp}$
			Cascade (xi)	$\Xi^0\ \bar{\Xi}^0$	1314.9	0	1/2	−2 +2	2.9×10^{-10}	$\Lambda\pi^0$
				$\Xi^-\ \bar{\Xi}^-$	1321.3	−1 +1	1/2	−2 +2	1.7×10^{-10}	$\Lambda\pi^{\mp}$
			Omega	$\Omega^-\ \bar{\Omega}^-$	1672.2	−1 +1	3/2	−3 +3	1.1×10^{-10}	$\Xi^0\pi^{\mp}$, $\Xi^{\mp}\pi^0$, $\Lambda K^{\mp}$

* 'Long-lived' here means with lifetime much greater than 10^{-23} s.
Antiparticle symbol is shown after particle (if different). Mass is in MeV, charge in units of the proton charge and lifetimes in seconds.

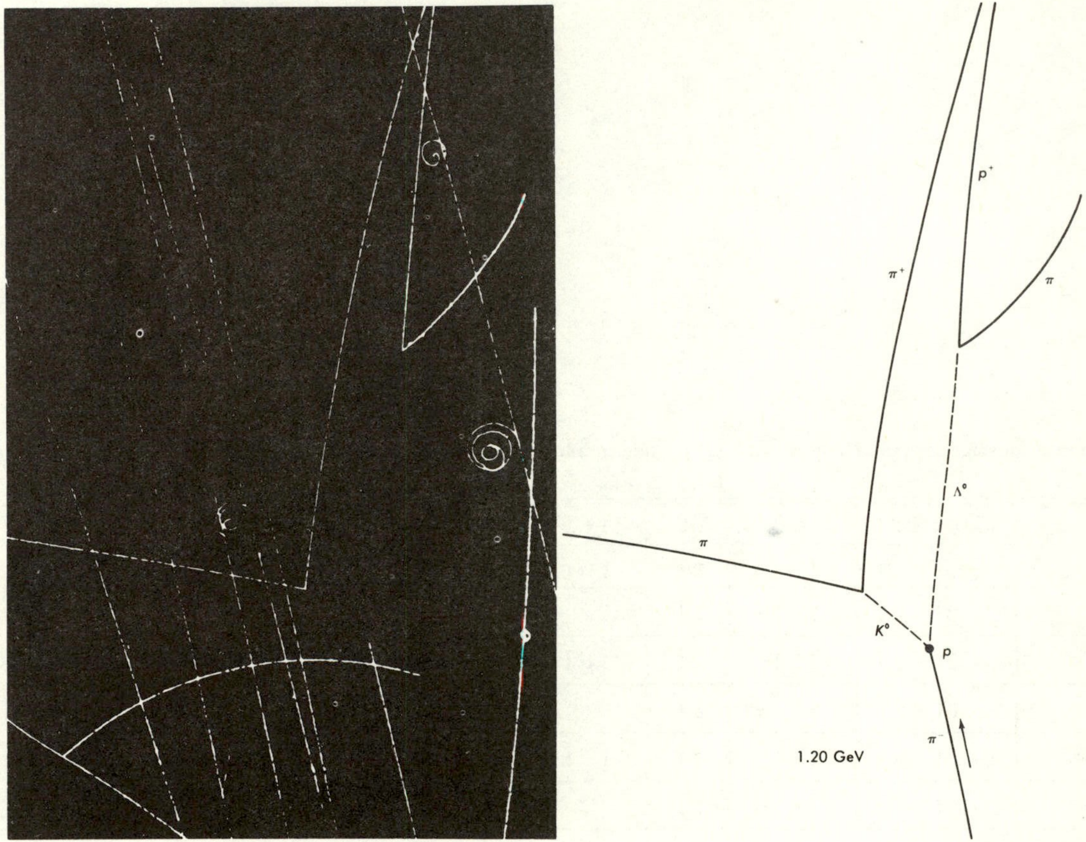

Fig. 5.6. Bubble chamber photograph of the production of strange particles. A high-energy pion (bottom right) collides with a proton in the chamber to give rise to two strange particles, Λ and K^0. Because they are electrically neutral they leave no tracks, but their decay products (p,π^- and $\pi^+\pi^-$) can be seen in the photograph.

Aside from the eta the properties of the new mesons and hyperons presented an immediate and profound puzzle, which earned them the name *strange particles*. It is easy to understand why they were considered strange. Take for example, the lambda hyperon. It may be produced in π^-p collisions, but it also *decays* into a pion and a proton

$$\pi^- + p \rightarrow \Lambda \rightarrow \pi^- + p$$

so on the face of it the Λ seems like a sort of temporary union of the π^- and p, like two insects that mate for a while then fly apart. Now there is a very deep law of physics, closely connected with time-reversal symmetry discussed in the previous section, which states that every process which goes one way can also go the other way – physics is *reversible* – so the behaviour of the lambda appears to accord with this expectation. The trouble is that the backward process $\Lambda \rightarrow \pi^- + p$ turns out to take about 10 million million times longer than the forward process $\pi^- + p \rightarrow \Lambda$, and that is very strange indeed.

The minuteness of the times involved masks the full significance of this enormous discrepancy. The forward process (Λ production) takes place in a mere 10^{-23} s. This is the time for light to cross the range of the nuclear force (10^{-15} m) and is typical of processes caused by the strong interaction. This force is so strong that it almost always produces a reaction if given enough time (contrast, say, the electromagnetic interaction which is only successful on one in every 137 occasions). The time needed is merely that required to inform the proton of the pion's proximity, information which, at the speed of light, takes just 10^{-23} s. On the other hand, although the decay of the lambda is simply the reverse process, also involving strongly interacting particles, it takes as long as 2.6×10^{-10} s for it to disintegrate. It is this relative longevity which enables a measurable (and invisible) track to occur in cloud chambers: the neutral lambda may travel a few centimetres in that time. In spite of the smallness of both the reaction times, the tardy decay of the lambda is as relatively peculiar as if a ball were thrown in the air and did not fall back down again for a million years.

The strange suspended animation is observed in all the hyperons and the kaons. An explanation was put forward in 1953 by the American physicist Murray Gell-Mann (Nobel physics prize 1969) and the Japanese physicist Kazuhiko Nishijima. Their idea is simple and elegant. If we ask, why does the proton have an infinite lifetime, the answer is: because of baryon conservation. The only thing that prevents it collapsing into a positron is the law of conservation of baryon number (see p. 159). This is an explanation, but not an understanding, because we do not know what baryon number is. Nevertheless, as a label it is useful in explaining other reactions too. Now the strange particles do not have *infinite* lifetimes, but in terms of 10^{-23} s time units, they live immensely long – 'almost' for ever. Therefore, Gell-Mann and Nishijima postulated that the strange particles carry a new type of label, which obeys yet another subatomic conservation law. An appropriate name for the label was chosen: *strangeness*. Like baryon and lepton number, strangeness is an invisible attribute. We cannot make a macroscopic strangeness field, or push a ball with a strangeness force. It is simply an internal property of the particle with no everyday counterpart. Nevertheless, strangeness is very real in the microworld, and regulates the behaviour of many particle processes.

The idea is that all the strange particles are endowed with whole number units of strangeness, rather like electric charge or baryon number. The K^+ and K^0 mesons have $+1$, Λ and all the sigmas -1, Ξ^- and Ξ^0 have -2 and Ω^- has -3 (see the table on pp. 172–3). Before explaining how the strangeness quality can account for the unusual behaviour of the new particles, some evidence for its reality should be given. The existence of the strangeness stuff on these particles is well illustrated by the trio of sigma hyperons, Σ^+, Σ^- and Σ^0. These are rather like the trio of pions π^+, π^- and π^0, which also come in charged and neutral varieties. The pions, though, have no strangeness. Now if strangeness on a particle is real, quantum field theory requires that its *antiparticle* should have equal and opposite strangeness. Experiments show that *all three* sigmas have the *same* strangeness of -1. Hence, there should exist *three other* sigmas with strangeness $+1$, the antisigmas (denoted $\bar{\Sigma}^+$, $\bar{\Sigma}^-$ and $\bar{\Sigma}^0$). This is indeed found to be so. The neutral antisigma, $\bar{\Sigma}^0$, is distinct physically from the neutral sigma, Σ^0; they do not enter into the same interactions. In the case of the pions, the absence of strangeness allows the π^0 to be its own antiparticle. Similarly the π^+ is the antiparticle of the π^-, but the Σ^+ is *not* the antiparticle of Σ^- (it even has a different mass). That role is reversed for $\bar{\Sigma}^-$. Similar remarks apply to the other strange particles. The eta meson (η) has zero strangeness and, like the photon and π^0, is its own antiparticle.

We now turn to the question of how strangeness explains the strange decay behaviour of the strange particles. According to Gell-Mann and Nishijima, in any reaction involving strong or electromagnetic forces, strangeness must be conserved, just like charge or spin. This law requires that strange particles are always created in pairs with equal and opposite strangeness numbers. The pairs need not be a particle–antiparticle combination but could, for example, be a K meson and a lambda

$$p + \pi^- \rightarrow K^0 + \Lambda.$$

Thus, the Λ is never produced on its own, but always in the company of another particle. Strangeness is conserved here because the ingredients, π^- and p, are not strange (have strangeness zero) while the K^0 has $+1$ and Λ has -1, which cancel.

The two electrically neutral strange particles fly off very fast in different directions, and as neither stands much chance of running into another strange particle which just happens to be lying around, they cannot decay back into particles like pions or protons without violating the law of strangeness conservation. Now all the conservation laws we have considered until now are supposed to be *absolute*, and obeyed by all four forces of nature. The novel departure in the case of strangeness is that, although conserved by strong and electromagnetic forces, the *weak force* can violate it. It is thus a *partial* conservation law. If there were only strong interactions in the world, the K mesons and hyperons would live for ever, as stable particles, and the world would be full of them. However, they are both coupled to the weak interaction as well, and eventually succeed in decaying by this route. Because of its relative feebleness, the weak decay takes from about 10^{13} to 10^{15} times longer to kill the hyperons and kaons than the strong force took to create them. The strange suspended animation is explained!

In the weak decay the Λ disintegrates into a nucleon and a pion: $\Lambda \rightarrow p + \pi^-$ or $\Lambda \rightarrow n + \pi^0$. The K mesons usually explode into muons or pions (e.g. $K^+ \rightarrow \mu^+ + \nu_\mu$, or $\pi^+ + \pi^0$, $K^0 \rightarrow \pi^+ + \pi^-$ or $2\pi^0$ together with many less frequent decay modes).

The number of possible reactions involving the other strange particles is enormous, though strangeness conservation forbids many of them. Various decay routes are summarized in the table on pp. 172–3: Note that all the hyperons eventually produce either a proton or a neutron. From this we may conclude that the hyperons must carry baryon number, to pass on to these products. They also carry intrinsic spin for the same reason. The K and eta mesons, on the other hand, have zero baryon number. Some strange particles undergo very complicated multi-stage disintegration. One example, the decay of the Ω^-, is shown in Fig. 5.7.

Strange particles are typically produced in proton–proton, or pion–nucleon collisions, for example

$$p + p \rightarrow \Xi^0 + p + K^0 + K^+$$

or by nucleon–strange particle collisions such as

$$\Lambda + p \rightarrow n + p + \tilde{K}^0$$

or

$$K^- + p \rightarrow K^+ + \Xi^- .$$

Many different reactions of this sort have been studied.

5.4 *Weak interaction revisited*

In Chapter 4 Fermi's model of a universal weak interaction was described, in which four fermions participate. Neutrinos were always involved somewhere in the reactions described there. In the previous

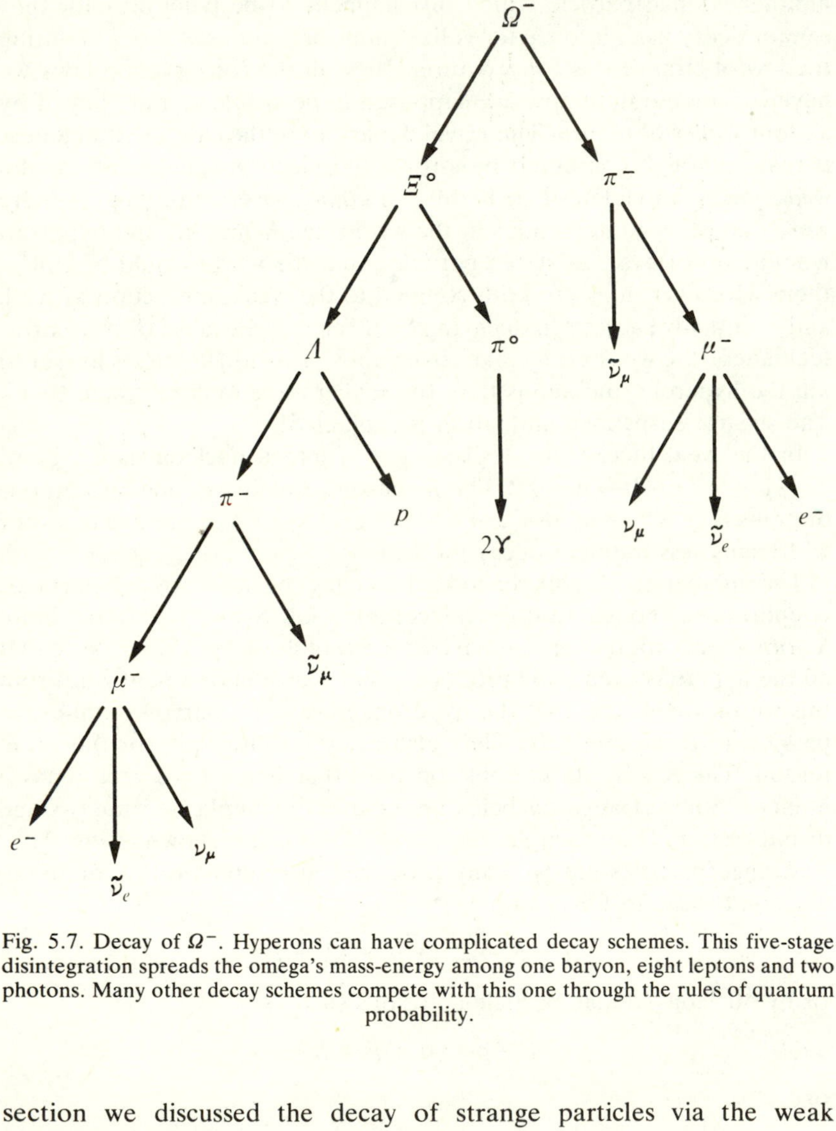

Fig. 5.7. Decay of Ω^-. Hyperons can have complicated decay schemes. This five-stage disintegration spreads the omega's mass-energy among one baryon, eight leptons and two photons. Many other decay schemes compete with this one through the rules of quantum probability.

section we discussed the decay of strange particles via the weak interaction. Typical of such processes is

$$\Lambda \rightarrow \pi^- + p.$$

In this reaction there are only two fermions (Λ and p) and no neutrinos! How can we relate such an interaction to Fermi's theory?

A similar situation occurs in the decay of the charged pion. Remember that the *neutral* pion decays *electromagnetically* into two gamma rays

$$\pi^0 \rightarrow 2\gamma$$

in about 10^{-16} s, whereas the charged pions decay into muons after more than 10^{-8} s

$$\pi^+ \rightarrow \mu^+ + \nu_\mu$$

$$\pi^- \rightarrow \mu^- + \bar{\nu}_\mu .$$

The much longer lifetime of the charged pion over its neutral counterpart indicates a much weaker force at work, and the presence of the neutrino points to the weak interaction. However, once again, only *two* rather than four fermions are apparently participating.

In spite of these superficial differences, it is possible to regard processes such as the decay of strange hyperons and the charged pion as examples of a universal four-fermion weak interaction by invoking the idea of virtual particles once more. Recall from p. 145 the argument for the decay of π^+ using the following possibility:

$$\underset{\text{strong}}{\pi^+ \rightarrow} \underset{\text{(virtual)}}{p + \bar{n}} \underset{\text{weak}}{\rightarrow} \mu^+ + \nu_\mu$$

where *four* fermions (p, $\bar{n}$, μ^+, ν_μ) take part in the latter, weak interaction, step. A similar argument may be used to explain the decay of the Λ hyperon by supposing that the pion is created initially as a $\bar{p}$, n pair. The four fermions are then Λ, p, $\bar{p}$ and n.

The decay of Λ still has a notable difference from the other weak processes discussed so far in that the four fermions involved are *all hadrons*. Note that, as with electric charge, weak charge can be carried by both hadrons and leptons. The study of weak interactions in hadrons is severely complicated by the fact that they are also subject to the strong interactions, which cannot be satisfactorily treated as a small perturbation. In contrast, weak interactions involving leptons are very well understood, at least at low energies. The electromagnetic interaction, with strength 1/137, is no problem and the weak force, being still weaker than this, can also be handled satisfactorily using perturbation approximation techniques.

Some weak processes exist in which all the four fermions are leptons. For example, the decay of the muon

$$\mu^- \rightarrow e^- + \nu_\mu + \bar{\nu}_e .$$

or

$$\mu^+ \rightarrow e^+ + \bar{\nu}_\mu + \nu_e \, .$$

Such reactions are admirable for studying the weak force uncluttered by strong or electromagnetic complications. The first question which comes to mind is: what is the nature of the weak current? How closely does it resemble the more familiar electric current? A related question is: how *strong* is the weak charge; what coupling constant plays the role of $e^2/\hbar c = 1/137$ and what is its value?

The decay of the muon can be carefully studied under controlled conditions by producing beams of pions in particle accelerators and allowing them to decay into muons through the reaction $\pi^- \rightarrow \mu^- + \bar{\nu}_\mu$, $\pi^+ \rightarrow \mu^+ + \nu_\mu$. Although the fate of the neutrinos cannot be directly monitored due to their exceedingly low tangibility, it is possible to measure both the direction of the muon's spin and the direction of motion of the emitted electron (or positron) when the muon decays. The result of this simple observation has the most profound consequences. It is found that although, for example, the electron can be emitted from μ^- over a whole range of angles relative to the muon's spin axis (depending on which way the neutrinos are emitted) there is a *preference* for the electron to fly off towards the side from which the spin of the muon appears *clockwise*, rather than the other way (see Fig. 5.8).

When this effect was first discovered early in 1957 (actually through the study of another weak process rather than muon decay) it caused consternation among physicists. The reason for the astonishment concerned the status of one of the fundamental symmetry principles of nature – the invariance of the laws of physics under space reflections (parity reversal) discussed in Section 5.1. There it was pointed out that an image of the world depicted in a mirror seems very possible. Before 1957 there was an almost universal assumption among physicists that the laws of physics would not be violated by a mirror world. However, they were wrong. Look at the mirror image of the decay of the muon (Fig. 5.8). When reflected in a mirror the spin of the muon is reversed from left-handed to right-handed rotation. In this picture the electron chooses the side from which the muon appears spinning *anti-clockwise* (our view in the mirror shown in Fig. 5.8). The mirror therefore alters the relationship between the spin direction and the direction of the emergent particle. What an extraordinary result this is. For some peculiar reason the electron likes to choose the clockwise direction to emerge in. It seems a near-impossible feat that such a simple entity as an electron is capable of distinguishing clockwise from anti-clockwise or right-handed from left-handed (some animals can't do this).

It is important to realize that the mirror image of muon decay is not simply unobserved, it is actually impossible – it violates the laws of the weak interaction. The significance of the reflection asymmetry is much more fundamental than the observation that all London buses have their cabs on the right, for there is nothing *unphysical* about a left-hand drive London bus, only that we don't ever encounter one. The muon decay asymmetry is a lopsided *law*, whose reverse image is actually *forbidden*.

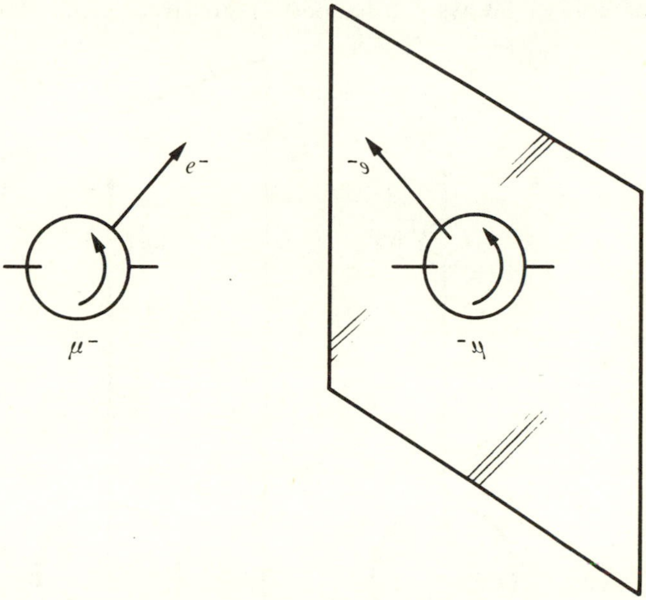

Fig. 5.8. Parity breakdown. When μ^- decays, the electron prefers to emerge towards the right side of the spin axis as shown, rather than the left (two neutrinos, not shown here, are also emitted). Clearly this lopsided tendency is asymmetric under reflections: the mirror image shows the electron choosing the left side. This is the behaviour adopted by the antiparticle decay.

The reflection asymmetry manifested by the weak interaction is apparently *not* shared by the laws governing the electromagnetic and strong interactions. A revealing comparison between the electromagnetic and weak forces concerns the nature of the photon versus the neutrino. As mentioned on p. 161 the neutrino is always found to spin like a *left*-handed corkscrew. No right-handed neutrinos exist. Reflection in a mirror converts a left hand to a right hand, so the mirror image of a neutrino is an impossible object in the real world. This already suggests that neutrino processes involve reflection-symmetry violation. In contrast, photons (whose spin is twice that of the neutrino) can rotate either left- or right-handedly, so the mirror image of one type of photon is an equally possible photon. In fact, left- and right-handed photons are none other than the quantum counterpart of left- and right-handed circularly polarized light or radio waves, familiar from ordinary optics and telecommunications.

These ideas are graphically illustrated by the decay of the pion. When the charged pion disintegrates, it produces a muon and a neutrino travelling in opposite directions with opposite spins. The neutrino from π^+ is, of course, always left-handed: there is no such thing as a

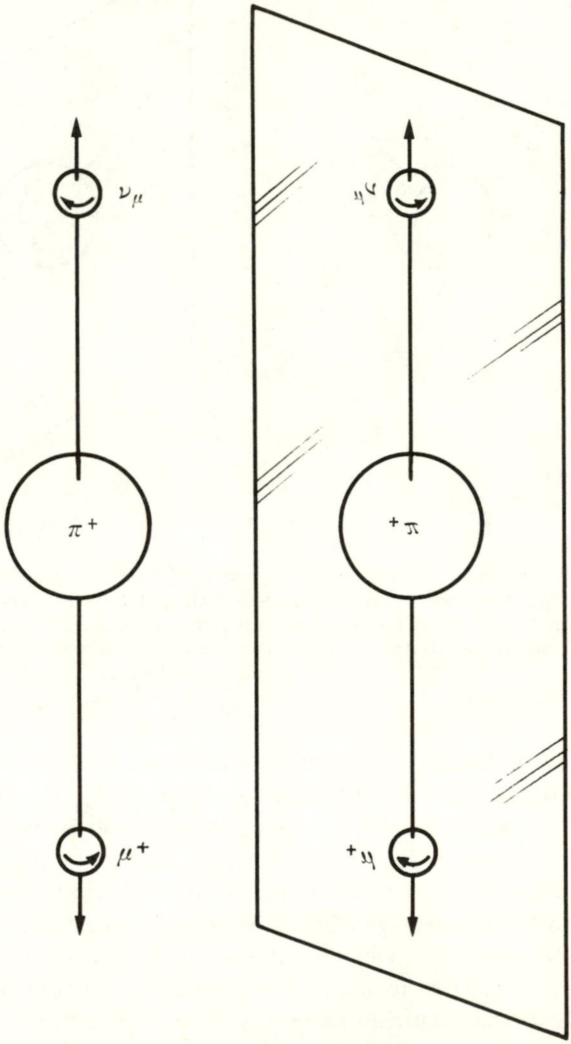

Fig. 5.9. The view of π^+ decay shown in the mirror would clearly be impossible in the real world as it involves a right-handed neutrino.

right-handed neutrino. Hence the mirror image shows an impossible situation (see Fig. 5.9). Similarly the antineutrino from π^- is always right-handed, and hence not reflection invariant. In contrast, the neutral pion, π^0, decays into two photons which have opposite rotation. Either one may be emitted in the 'up' direction, so the mirror image (see Fig. 5.10) is equally probable.

In Section 5.2 it was pointed out that the PCT theorem requires that, even though reflection symmetry (P) is violated, the combined operation PCT should leave the laws of physics invariant. If the decay of the μ^+ is studied it is found that, unlike the electron from μ^-, the positron is emitted preferentially in the direction of *anti-clockwise* rotation of the spin of the muon. This equally asymmetric characteristic is therefore the mirror image of the μ^- decay. It follows that although the reflected image shown in Fig. 5.8 of the electron emission from μ^- is an impossible process, the *charge-reversed* (C) reflection is quite all right, being just what is observed when a μ^+ emits a positron. Thus, under the combined symmetry operation CP, the weak interaction seems invariant. We can conclude from this that these processes are also symmetric under time reversal (T), otherwise PCT would be violated.

The discovery of reflection asymmetry in weak interactions means that the weak charge cannot be precisely analogous to electric charge. It was mentioned on p. 168 that electric charge and its associated field are reflection symmetric, but a magnetic charge, if it existed, would not be. It is necessary to consider the weak current to be analogous to a mixture of electric and magnetic currents in order to incorporate the observed asymmetry.

Assuming a current of this form it is possible to calculate the rate of muon decay from the mathematical theory of quantum mechanics. The answer depends on the *strength* of the weak charges, so that a measurement of muon decay provides a value for the fundamental unit of weak charge carried by the muon and other particles. If we call the weak charge g, then the quantity $g^2/\hbar c$ is the analogue of the fine structure constant $e^2/\hbar c$. It so happens that data concerning the muon decay rate and other weak processes studied so far can only give the value of the ratio g/M_W where M_W is the mass of the intermediate W particle. As this is as yet unknown we cannot fix $g^2\ \hbar c$ definitely. However, the *effective* interaction is actually $(M/M_w)^2\, g^2$, where M is about the same as a proton mass. One then obtains

$$(M/M_w)^2\, g^2/\hbar c \simeq 0.57 \times 10^{-6}$$

which is more than 12 thousand times weaker than the electromagnetic coupling (1/137).

When hadrons are present in weak interactions the situation is complicated by the presence of strong interactions which may even swamp the weak processes completely. If the assumption is made that the same basic type of interaction which disintegrates the muon and operates between lepton currents also operates when hadron weak currents are present then some successful predictions can be made in spite of strong force complications. One feature which apparently persists when hadrons are involved is the reflection asymmetry. In fact,

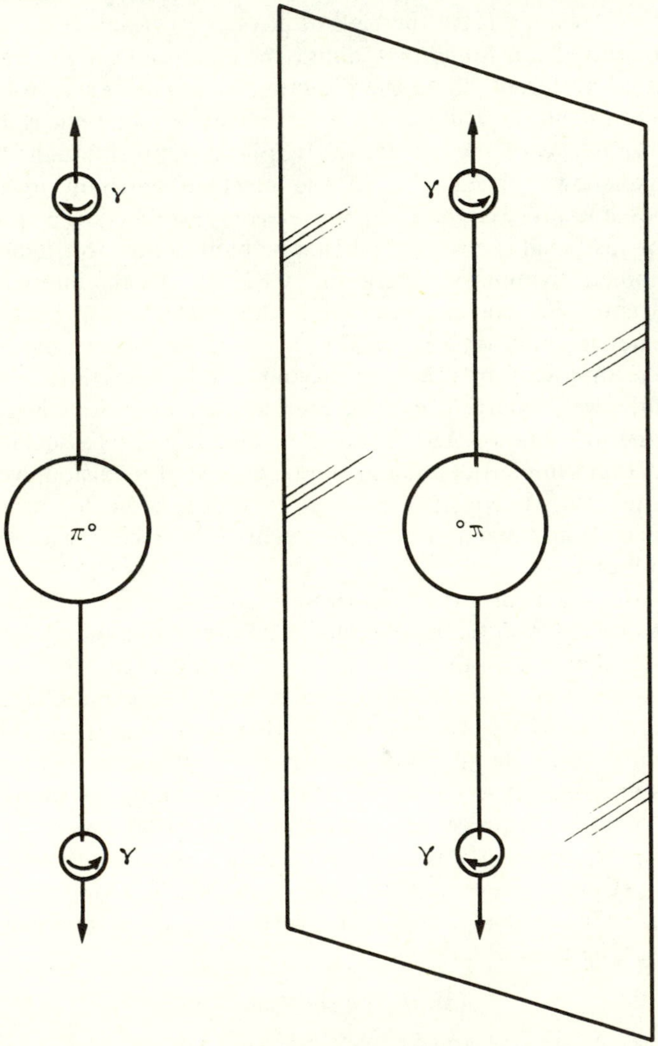

Fig. 5.10. The view of π^0 decay shown in the mirror is as equally acceptable as the original, because photons may spin either way. Electromagnetic forces do not violate mirror symmetry.

the first experiment to confirm its existence was carried out using the beta decay of radioactive cobalt. The experiment, performed in 1957 by the Chinese American, Mrs Chien-Shiung Wu, detected the preferential emission of beta rays (electrons) in the direction away from the spin vector direction of the cobalt nuclei. To achieve this observation, Mrs Wu had to arrange for many cobalt nuclei to line up their spins in a prearranged direction and then measure the angular distribution of the emitted electrons. The result confirmed the 1956 prediction of the two Chinese Americans Tsung Dao Lee and Chen Ning Yang that reflection symmetry might be violated in weak interactions. In 1957 Lee and Yang shared a Nobel prize for this remarkable insight.

One curious puzzle presented by hadrons concerns the weak interaction decay rates. In Section 5.3 it was explained how the disintegration of strange particles is controlled by the weak force. When the rates (or lifetimes) of strange hadron decays are compared with similar decays of non-strange hadrons, there is a large discrepancy. For example, the strange hyperon Λ can decay as follows

$$\Lambda \rightarrow p + e^- + \bar{\nu}_e$$

which has the same end product as the decay of the neutron

$$n \rightarrow p + e^- + \bar{\nu}_e \,.$$

Yet a comparison of the lifetimes of the two processes, corrected for effects of energy differences, shows that the intrinsic decay rate of Λ is some 17 times slower than that of the neutron. It is as though the change in strangeness necessary to convert the Λ into its non-strange progeny is in some way inhibiting the decay process. This can be understood by supposing that the weak current of hadrons divides into two components, one which stimulates the strangeness conserving decays, the other causing the strangeness-changing decays. Measurements indicate that the former is about five times as strong as the latter, which accounts for the relative feebleness of the decay of the Λ and other strange particles.

We come now to the question of the absolute strength of the weak charge on hadrons. Because of the strong interaction, a particle such as a proton is dressed in a cloud of virtual mesons which will have their own weak charges. There is no *a priori* reason to suppose that the combined weak charges of the proton and all its attendant mesons will turn out to be the same as that of the weak charge on leptons. The electron, for example, does not have a meson cloud around it. So we have no reason to expect that $g^2/\hbar c$ will be the same for hadrons and leptons. Remarkably enough, though, the values are very close, particularly for the reflection symmetric part of the current.

This observation has an important implication for the analogy between electric and weak charge. One of the fundamental laws of electricity is that electric charge cannot be created or destroyed. In all processes it is conserved; if one particle changes into another, the same units of electricity are always handed on to the new particles. Electricity never appears in, or disappears from, the universe. As a result of electric charge conservation, whenever a charged particle is created, whether real or virtual, it must be accompanied by another particle of equal and opposite charge. Thus, for example, the meson cloud which surrounds a neutron must contain at any time an equal number of positively and negatively charged mesons. This fact will ensure that the *total* charge of the neutron is still zero, even though it has a cloud of electricity in its proximity. Similarly, the total charge on a proton is unaffected by the presence of its own charged meson cloud, which explains why the electric charge on leptons is the same as that on hadrons (e.g. the positron and proton both possess exactly one unit of positive charge). The question now arises as to whether *weak* charge is likewise conserved. By analogy with the electric case, the equality of g on leptons and hadrons seems to be good evidence that this is so, at least for the reflection-symmetric component.

We end this section with some comments on the decay of the neutral K meson through the weak interaction. It has been pointed out that although the photon, graviton, π^0 and η mesons can be identified with their own antiparticles, the K^0 meson cannot, because it has strangeness +1. Thus, its antiparticle, $\tilde{K}^0$, must have strangeness –1. As far as strong and electromagnetic interactions are concerned, the K^0 and $\tilde{K}^0$ are distinct. However, the weak interaction is strangeness-blind, and cannot distinguish them. The weak force is responsible, of course, for disintegrating both of them into non-strange real particles. For example

$$K^0 \rightarrow \pi^+ + \pi^-$$
$$\tilde{K}^0 \rightarrow \pi^+ + \pi^- .$$

As the products are the same, there is clearly the possibility of

$$K^0 \rightarrow \underset{\text{(virtual)}}{\pi^+ + \pi^-} \rightarrow \tilde{K}^0$$

with virtual intermediate pions. Indeed, the neutral K meson must be considered to oscillate, under the stimulus of the weak force, back and forth between K^0 and $\tilde{K}^0$, living part of its life as each.

Subtleties occur when this 'dynamic duo' decays. The point is that K mesons, like all subatomic particles, are really to be described by *wave* motions, in accordance with the ideas of quantum mechanics. Recalling the discussions of Section 2.5, when waves are superimposed, *interference* effects become important. The waves either reinforce or cancel depending on whether they are in or out of phase. When the two faces of the neutral K meson are superimposed, the phases of their waves must be taken into account. It can be shown that they combine together in two distinct ways: in phase, and exactly out of phase. This result would be of little importance if it were not for the fact that the *charge-reversed mirror image* of the latter combination is different, whereas that of the former is the *same*. When the neutral K meson decays it wants to break into two pions. However, the only quantum wave consisting of two pions is charge and space reflection symmetric. It follows that the out-of-phase component, usually called K_2^0, cannot decay into two pions without violating CP invariance. On the other hand, the former component, called K_1^0, can decay in this manner.

These observations have the bizarre implication that only 'half' the K meson, K_1^0, can decay this way. What happens to the other half? It decays into *three* pions ($K_2^0 \rightarrow \pi^+ + \pi^- + \pi^0$ or $3\pi^0$) which do *not* have a reflection symmetric wave pattern. Moreover, because of the additional energy tied up in the extra pion, the three-particle decay is some thousand times slower. We thus have the extraordinary image of the K meson as a sort of hybrid particle. It can either be in the K_1^0 condition and decay into two pions after about 10^{-10} s, or in the K_2^0 condition and disintegrate into three pions after about 10^{-7} s.

Suppose we start out with a beam of K^0s. After a few times 10^{-10} s all the K_1^0 component will have disappeared into pairs of pions, leaving pure K_2^0 further downstream. This is an equal mixture of K^0 and $\tilde{K}^0$, so if the beam encounters ordinary matter, the $\tilde{K}^0$ will get absorbed by the protons in the reaction

$$\tilde{K}^0 + p \rightarrow \Lambda + \pi^+$$

leaving pure K^0 once again! In this way the beam of K^0 mesons can be repurified even though decay into pion pairs has occurred.

Amazing though these properties of K^0 are, a bigger surprise is in store. In 1964 it was discovered by two groups of physicists, one group working at Princeton, the other at Illinois, that on about two in every thousand occasions, K_2^0 decays into *two* pions! The conclusion, as shattering as the discovery of the collapse of reflection symmetry, is that *CP symmetry is violated*. Either the PCT theorem is wrong, and with it the whole basis of relativistic quantum field theory, or time-reversal symmetry is violated, too. Careful experiment showed the latter to be the case. The decay of the neutral K meson can distinguish between past and future, a result more astonishing than its ability to distinguish left from right. As remarked on p. 157, the asymmetry in time observed in the everyday world is a consequence of the *macroscopic organization* of vast numbers of particles and fields. When we follow the progress of a clock running down we are witnessing an activity that involves countless billions of atoms behaving collectively. What possible comparable mechanism can exist in the subatomic world to enable the K^0 meson to tell which way in time clocks run down and people grow old? The explanation for this feat remains a mystery. No other particle seems to share the ability.

6

Unity or diversity?

In the previous chapter it was explained how total anarchy in the microworld is at least partially averted by the rule of law. Conservation of energy, momentum, spin, electric charge, lepton and baryon number and, at least for hadronic reactions, strangeness, help to order the *behaviour* of the inhabitants of the subatomic zoo. In spite of this, no law has been discovered which limits the *number* of species. The discovery of new types of subatomic particles did not stop with the strange ones, and continues today almost monthly.

The search for tidiness proceeds in three ways. First, new conservation laws are sought which further limit the behaviour of all the particles. Some of these will be discussed in the next sections. Secondly, enormous advances have been made since the early 1960s in unifying the proliferation of particles into *families*, and understanding these groupings as a manifestation of some kind of underlying building blocks – 'atoms' of elementary matter. Thirdly, equally exciting and significant progress has been made recently in understanding the *forces* of nature.

Let us summarize what has been said so far about the four forces. Their respective strengths are roughly in the ratios 10^{-40}: 10^{-6}: 10^{-2}: 1. Gravity, the weakest is, from the subatomic point of view, totally mysterious because of the complete absence of any observable quantum effects. However, according to the most fundamental physical principles, which are built into our present theory of gravity (Einstein's general relativity), *all* matter is equally subject to the gravitational force. Being universal, gravity is in a sense the most important force of nature. Every subatomic particle (even the graviton) is caught in its grip. The reason for the extreme weakness of gravity is unknown, though it has been argued that if it were not so the universe would have collapsed long before scientists had evolved to worry about the issue.

Next in strength comes the weak interaction, distinguished by its short range, and its disregard of the laws of parity, CP and strangeness conservation. Its extreme weakness compared to the remaining two forces seems a puzzle.

Electromagnetism is the best understood and well behaved of all the forces. It has familiar consequences in daily life, and appears to operate in the microworld according to the same laws and principles (when quantized) as the macroscopic world. The only real mysteries concern the reason for the absence of magnetic monopoles and for the value of 1/137 for the fine structure constant ($e^2/\hbar c$).

The strong interaction is the least well understood of all the forces. The original hope of Yukawa that internucleon forces could be simply described by the exchange of a single meson field quantum has been dashed. The existence of a whole army of mesons more massive than the pion means at best that the short-ranged part of the nuclear force field is extremely complicated, depending on an elaborate network of particle exchanges. At worst it appears that the nuclear force is not really elementary at all, but a little understood vestige of a more fundamental *internal* hadronic interaction, rather as chemical bonds between atoms are vestiges of the electromagnetic forces which bind the electrons to the nucleus.

We are moving towards a picture with three distinct types of particles. There are the leptons, apparently structureless and elementary, interacting weakly with each other and with different particle types. Then we have the hadrons that are altogether more complex entities, showing every sign of possessing an *internal* structure and a finite size (about 1 fm). They are presumably held together by a powerful version of the strong force. Finally come the graviton and photon, in a strange little class apart.

Physicists are bound to enquire: why four forces? What is the relation between them? Why are their strengths so disparate?

The history of theoretical physics is the history of *unification*. The models constructed using mathematics derive their power by relating together previously unrelated phenomena. The unification of the forces of nature into one single, all-embracing interaction is surely the ultimate goal of subatomic physics. Yet the task seems a daunting one, especially when the widely different properties of the forces are considered. In spite of this it has recently been shown how a mathematical framework does exist for combining together at least two of the four forces into a unified field theory. Moreover, some experimental verification of this idea is starting to appear. There is still a long way to go, but it is clear that the nature of the forces and of the subatomic particles must go hand in hand. No full understanding of one will come without a full understanding of the other.

6.1 *More particles*

We have come a long way from the electrons and protons of Thomson, Rutherford and their colleagues. The number of particles discussed so far has reached 20 (plus their antiparticles). In the bizarre perspective of subatomic physics, they are all 'long-lived'. The proton, electron, graviton, photon and neutrinos are all apparently completely stable against decay. The remainder decay only weakly, through the weak interaction in fact, except for the neutral particles Σ^0, π^0 and η, which are destroyed by the electromagnetic force. In the case of particles which can disintegrate into hadrons, decay under the much more rapid strong interaction is inhibited by strangeness conservation (except for Σ^0 and η, which will be discussed in Section 6.2).

The question which now comes to mind is whether there exist *non-strange* massive hadrons that can use the strong force to decay. Such particles would only last for around 10^{-23} to 10^{-24} s before disintegrating into lighter hadrons, and this duration is too brief for them ever to be detected directly. They live and die within the space of a single atom. However, their existence might show up in scattering experiments in which the short-lived hadron is created during the collision of other hadrons. When it decays, it will spew its progeny out in certain well-defined directions. To the experimenter, who is unable to witness the fleeting formation and demise of the new hadron, the event will appear as a dramatic increase in the scattering power of the target hadrons.

To create new, heavy hadrons, enough kinetic energy must be given to the colliding particles to supply the rest mass of the created particle. However, because its lifetime is so short, the energy requirement is actually rather flexible owing to the Heisenberg uncertainty principle, Equation (2.8). Indeed, a hadron with a lifetime as short as 10^{-23} s can borrow up to hundreds of MeV energy for its brief duration. Consequently, as the experimenter turns up the energy input, what he sees is not the sudden onset of a reaction as soon as the rest mass energy is reached, but rather a slow build up to a maximum reaction, followed by an equally slow decline (see Fig. 6.1). This type of damped out response is similar to the phenomenon of resonance (indeed, taking into account the wave nature of matter, it *is* a resonance) so the short-lived particles are called *resonances*.

The first resonance to be observed was found by Fermi in 1952 from cyclotron experiments in Chicago, in which pions were scattered by nucleons. For example, when the incoming pion energy is about 190 MeV the scattering cross-section of protons rises conspicuously showing a peak with a width of about 120 MeV. The explanation is that the pion and proton combine to form a new hadron, called Δ, with a

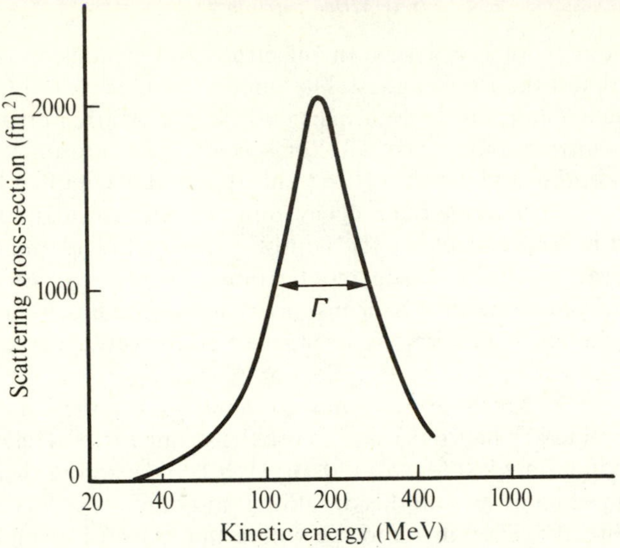

Fig. 6.1. Resonance. When π scatters from p a particle, or resonance, Δ may form around 190 MeV pion energy. The width Γ of the resonance is a measure of the lifetime of Δ.

mass of 1236 MeV and a spin of three halves. After 10^{-23} s Δ disintegrates into the proton and pion once more. For example

$$p + \pi^+ \rightarrow \Delta^{++} \rightarrow p + \pi^+.$$

Note that, as positively charged pions are used, the Δ carries a *double* charge (++) to conserve electricity. Use of other pions, and neutrons instead of protons, will produce similar resonances with charges +1, 0 and −1.

In a way, Δ^{++} is rather like an excited state of the proton. Just as an atom can absorb a photon and become excited, so here the proton seems to absorb a pion, then emit it again. Moreover, p has spin one half, π^+ has zero spin, but Δ^{++} has spin three halves. We must therefore conclude that to generate the spin of the Δ, it would appear necessary for the temporary union of p and π^+ to have some sort of internal *orbital* angular momentum of one to add to the available spin of one half, making three halves in all.

Although there is no doubt that Δ exists as an independent entity, it may not seem clear that it represents a truly new subatomic particle. When an atom A absorbs a photon, resides temporarily in an excited state A*, then decays to A by emitting another photon, we do not think of A* as a new particle, distinct from A. This is in spite of the fact that A* is about 10^{-7}% heavier than A due to the additional mass-energy of excitation ($E = mc^2$ again). However, in the case of Δ, the excitation energy is as much as 160 MeV, or 15% of the combined proton–pion rest mass, and it seems appropriate to regard Δ as a completely autonomous entity in its own right.

In the years that followed Fermi's discovery many other excited states of the nuclear particles were found. In addition, a resonance between two pions was detected, with a mass of 770 MeV and a spin of one. It is called the ρ meson (rho meson). Virtual ρ mesons form in the cloud of virtual mesons that surround protons and so the existence of the particle affects the scattering of electrons from this cloud. Other, more massive resonances involving pions exist. They are the ω (omega) and f mesons. The latter can also be regarded as an excited state of the η meson. It is distinguished by having a spin of *two* units, like the graviton, a fact which we shall return to shortly.

All the above resonances have zero strangeness. Short-lived excited states of strange particles can also occur if they can decay via the strong interaction into other strange particles: many such cases are known. To date, the total list of all these resonant particles runs into hundreds, and there is no sign of any limit.

A closer look at the ρ meson reveals it to have the same labels as the photon: spin one unit, zero charge, zero strangeness, zero baryon and lepton numbers. In fact, it differs from the photon only in its mass. In short, the rho meson is a heavy photon. It follows that the only conservation laws that can prevent a photon turning into a ρ meson and vice versa are the laws of energy and momentum conservation. However, for a short while these can be suspended because of the uncertainty principle, allowing a photon to exist part of the time as a virtual ρ meson. The greater the photon energy, the less the borrowed amount, so the longer the fraction of time it spends as a rho meson.

The importance of this virtual transmutation is that unlike the photon, the ρ meson is a *hadron* – it feels the strong force. Thus, when the electromagnetic field couples to hadrons, it does not do so merely by acting on their electric charges, it also couples via the rho through the strong interaction. A closer inspection shows that two other mesons, the neutral φ and the neutral ω, also have spin one, and can act as intermediaries between photons and hadrons.

Experiments confirm these expectations. If very high-energy gamma ray photons are shot at neutrons and protons it is found that they are absorbed equally efficiently by *both* these nucleons, indicating that it is their common hadronness rather than the electric charge of the proton which is responsible for the interaction. If photons are scattered from whole nuclei, a similar phenomenon occurs. At low energies, when the virtual meson part of the photons are short-lived and ineffectual, it is found that the photons scatter in proportion to the total number of nuclear particles present. This is the situation expected if the photons see only the electric charges present. Because the electromagnetic force is so weak, the overwhelming number of photons travels right through the nuclei, with trajectories that are deviated but not stopped.

In contrast, at high energies the meson component of the photons becomes evident. Some of the photons will be mesons when they hit the nuclear surface. The strong hadronic interaction prevents the mesons from travelling more than a couple of fm into the body of the nucleus, so the amount of scatter that these particles suffer will be proportional, not to the total nucleon particle content, but to the surface area of the nucleus. The surface area is only proportional to the *two-thirds* power of the number so the contribution from surface scattering rises more slowly with the nuclear content than the contribution from scattering in the nuclear interior. At high energies, when the surface contribution is relatively more important, one does indeed find that heavy nuclei scatter high-energy photons somewhat less than in proportion to their total particle number.

In an ingenious extension of these ideas the Pakistani physicist Abdus Salam of London University and the International Centre for Theoretical Physics in Trieste, Italy, has suggested that gravitons also interact with hadrons through the intermediary of various spin-two neutral mesons, such as the *f* meson. Salam and his collaborators have used this to construct a theory of 'strong' gravity, involving mesons which satisfy a hadronic counterpart of the Einstein equations, complete with 'space–time' curvature, black holes and micro-universes!

If there was ever a hope that simplicity lay at the heart of matter, it appears to be utterly discredited. The variety and complexity of subatomic particles appears completely bewildering. Why do so many species exist in the subatomic zoo? How can we make sense of all these entities – or even order them in a meaningful way? In spite of strenuous efforts, nobody knows the answers to these questions. Only glimpses of a higher pattern have begun to emerge.

Some attempts to answer the questions have established fascinating and undoubtedly important new perspectives on the workings of nature, and if nothing else testify to the versatility and ingenuity of the human mind. Unfortunately they provide only shadowy shapes of some dimly perceived theory of matter. Because much of these modern developments require a knowledge of advanced mathematics, we can only give here something of the flavour of a few of them. The exposition is necessarily unbalanced because it favours those ideas that can be roughly described in non-mathematical terms. The important theory of Regge poles, for example, in which complex numbers are applied to angular momentum, cannot even be sketched in this book.

6.2 *Elementary particles?*

This Ariyan Eightfold Path; that is to say: Right view, right aim, right speech, right action, right living, right effort, right mindfulness, right contemplation.
(Buddha, Fifth century B.C.)

When describing the resonances, it was pointed out that an excited atom bears the same *qualitative* relationship to an unexcited one as the Δ particle does to a nuclear particle. We only choose to describe the atom as two different states of the *same* particle because the energy difference in the two states is such a small fraction of the total mass-energy. This suggests that when we encounter particles with only small mass differences separated from other particles by a large gap, we regard them as really the same particle, existing in two or more different states.

A good example of this is provided by the neutron and the proton. Their mass difference is only about 0.1% of their total mass, and no other particles exist with a mass anywhere near or close to n or p. The neutron and proton form an isolated family. Why do we then think of them as *different* particles? The obvious answer is that the proton is electrically charged, but the neutron is not. Certainly the charge does lead to some differences in properties when electromagnetic processes occur, but the strong interaction, which both n and p enjoy, completely swamps the puny electromagnetic force. As far as their identity as *hadrons* is concerned, there is little to choose between them. The strong force reactions of one are very much like those of the other. In this respect, then, the electric charge behaves more like a *label* than something which affects *behaviour*.

Physicists now think of the neutron and proton as really two states of the same particle, called the nucleon and denoted by N. The charge label distinguishes the two available states, n and p. At first sight this idea seems a fraud, but it becomes more justified on closer inspection. It was mentioned in Section 2.7 that a particle with intrinsic spin one half can have the spin axis pointing in only one of two directions: 'up' or 'down'. That is, a free particle with spin has only two available spin states. Suppose the particle is a proton. We know the proton is a tiny magnet, so if we place it in a magnetic field it will try to twist N–S. Because of the above-mentioned restriction on orientations, it either lies N–S or S–N. The magnetic energy will be different for the two different states (the proton has a natural tendency to seek the lowest energy state, but in a whole group not all will be able to divest themselves of the excess energy at the same time).

In a magnetic field therefore, a proton can have two slightly different masses, according to whether the spin points N–S or S–N. We could, of course, regard these two states as two different particles and say that the spin direction labels the particles. This is very similar to using the charge to label proton and neutron, and has encouraged the terminology *isotopic spin* to denote it (isotopic, because isotopes differ in their content of neutrons and protons). Although its labelling characteristics make it resemble ordinary spin, isotopic spin must not be imagined as some weird *mechanical* property. It has nothing to do with ordinary spin. The space in which the isotopic spin direction points is a totally fictitious, purely mathematical one.

Using the concept of isotopic spin, we can find other groupings of subatomic particles. The three pions π^+, π^- and π^0 have very nearly equal mass and can be regarded as a single pion whose isotopic spin is *one* unit, and can point in *three* (fictitious!) directions, one for each pion. The two K mesons, K^+ and K^0, can be treated as a doublet as can the two cascade particles, Ξ^- and Ξ^0. Similarly the three sigma hyperons Σ^+, Σ^-, Σ^0 form a triplet. The Δ resonance, with four charge states (+2, +1, 0, −1) has isotopic spin three halves.

If it were only a means of ordering particles into groups, the concept of iosotopic spin would have limited appeal. Its significance lies in something deeper. As remarked, the electromagnetic interaction regards n and p as different because it couples to electric charge. But in hadron processes the *strong* force dominates, and in all experiments so far it has been found that the strong force is 'charge-blind'. It treats the proton the same as the neutron, the three pions equally, and so on.

Some straightforward evidence of the charge independence of the strong interaction comes from examining the masses of certain light nuclei known as mirror nuclei, discussed briefly on p. 131. The example given there is the isotope of helium with one neutron and two protons, which is a sort of mirror twin to tritium – the isotope of hydrogen with one proton and two neutrons. These two nuclei differ only by the interchange of a single proton and neutron. A comparison of their binding energies (allowing for the additional electric repulsion between the two protons in the helium nucleus) shows that the strong force between proton–proton, neutron–neutron and neutron–proton are the same. Electric charge apparently plays no role in the strong interaction.

A fancy way of expressing the charge independence of the strong interaction is to say that it is invariant under rotations in isotopic spin space. What does this mean? If a system is invariant under rotations in ordinary space, it means that it does not matter in which direction the system is oriented, the forces are just the same. Similarly for isotopic spin, the strong force does not distinguish the 'up' orientation (proton) from 'down' (neutron). Naturally this is not true of the electromagnetic force, so electromagnetism destroys the isotropy (rotational invariance) of isotopic spin space, in the same way that a magnetic field destroys the isotropy of ordinary space, and introduces an energy gap between protons pointing up and down. In the isotopic spin case, this energy gap is manifested as the small mass difference between the neutron and the proton, which is presumably due to some internal electromagnetic coupling. The fact that the mass difference is such a small fraction of the total mass is evidence that the rotational symmetry of isotopic spin space is only weakly broken.

The rotational invariance of the strong force in isotopic spin space enables previously distinct hadronic processes to be considered as basically the same. Consider, for example, the interaction between pions and nucleons, $\pi + N$. The isotopic spin of π is one, that of N is one half. They can combine together to give a total of three halves or one half depending on whether they are parallel or antiparallel in the fictitious space. In the three halves state the combination describes $\pi^+ + p$ if all the spin points 'up' and $\pi^- + n$ if it points 'down' (there are also intermediate states describing $\pi^+ + n$, for example). Now the strong interaction cannot read the orientation of isotopic spin, which means that so long as the total isotopic spin of the $\pi + N$ system is the same (three halves) the strong force cannot tell which 'direction' it points in. Thus the interaction between π^+ and p is identical to that of $\pi^- + n$, as far as strong forces are concerned.

In Section 5.1 a connection was established between symmetries and conservation laws. This idea extends to the rotational symmetry of isotopic spin space, and predicts that, as far as hadronic processes are concerned, isotopic spin will be *conserved* (just as ordinary spin is conserved because of the rotational invariance of ordinary space). Consider, for example, the decay of the Δ resonance

$$\Delta^{++} \rightarrow p + \pi^+.$$

Both sides have isotopic spin three halves, which is therefore conserved in the decay.

Similarly, in the production of hyperons by proton–proton collisions

$$p + p \rightarrow p + \Lambda + K^+$$

the isotopic spins of the protons and the kaon are one half, while that of Λ is zero. Thus each side of the reaction has one unit of isotopic spin, as required for conservation.

These examples show that isotopic spin conservation is a much more powerful law than mere charge conservation (which it embodies). By ordering groups of particles with different charge into multiplets with similar hadronic behaviour, it leads to a great simplification in the variety of subatomic activity. However, it should come as no surprise that the weak interaction, which breaks strangeness, parity and time-reversal symmetry, also breaks isotopic spin symmetry. When the strange hyperon Λ decays, it does so via the *weak* interaction:

$$\Lambda \rightarrow p + \pi^- \text{ or } n + \pi^0 .$$

The initial isotopic spin is zero, but that of the final states cannot be.

As mentioned already the electromagnetic interaction also breaks isotopic spin symmetry, because it is not charge-blind. Even in the electromagnetic decay of the electrically *neutral* particles, the symmetry can be broken. We now have at hand the long-awaited explanation for the longevity of the non-strange η meson. Recall that this particle is a hadron that can decay into lighter hadrons:

$$\eta \rightarrow \pi^+ + \pi^- + \pi^0 \text{ or } 3\pi^0$$

and so there appears to be no obvious reason why it should not decay via the strong interaction in about 10^{-23} s. Instead it takes about 10^{-19} s. For comparison, strange particles decay via the weak interaction in about 10^{-10} s. This indicates a force mid-way in strength between the weak and strong, which points to electromagnetism, a guess confirmed by the fact that the decay products can also be gamma rays, rather than pions. Why doesn't η decay hadronically? Strangeness conservation, which prevents the strong decay of the strange particles, is no help as η has zero strangeness. The explanation lies with isotopic spin conservation. The η has zero spin and isotopic spin, but three pions cannot combine their motions and isotopic spins to give zero in such a way as to preserve the space and charge reflection properties of η. Hence isotopic spin is not conserved in the decay, so it cannot proceed under the strong interaction. The electromagnetic force, on the other hand, can break this conservation law, but its relative weakness results in a much slower decay. Notice from the table on pp. 172–3 that the neutral strange sigma hyperon, Σ^0, can use the electromagnetic route to decay also (note its very short lifetime for a strange particle). This is because it can thereby reach the Λ hyperon which has the same strangeness as Σ^0. Thus strangeness is still conserved in this decay. Isotopic spin is not.

Accepting the idea of isotopic spin introduces a small element of tidiness into the chaotic subatomic zoo. Some apparently distinct species are seen to be merely superficially different varieties of the same animal. If one imagines that in some magic way the electromagnetic interaction could be switched off, then the small mass splittings between the members of the isotopic spin species would disappear. The neutron and proton, for example, would become indistinguishable. So would the *K* mesons. There is, however, a proviso. Because the weak interaction can also see the direction of isotopic spin, it too will cause a small mass difference between members of an isotopic spin multiplet, though its relative feebleness will result in much smaller splitting than the electromagnetic interaction. Indeed, a careful comparison of the K_1^0 meson with its counterpart K_2^0 reveals a tiny mass splitting of around 10^{-11} MeV. (The electromagnetic mass splitting is, of course, absent between these neutral particles, though for comparison we can note that the splitting between K^0 and K^+ is 4 MeV.) If both electromagnetic and weak interactions are switched off, the laws of physics would have a new symmetry – isotopic spin rotation invariance – and many otherwise distinguishable particles would collapse together into the same animal.

Nothing has yet been said about how the electromagnetic and weak interactions are supposed to cause mass splittings. It must clearly come about through some internal mechanism whereby a particle can *act on itself* via these forces. The way in which this happens will be discussed further in the next section, but for now we simply note that the internal electromagnetic and weak action will obviously depend on whether we are dealing with a point-like particle, or something with a more complicated internal structure. Recall that, by considering a nucleus as a composite, structured body, it is possible to compute the electric energy, and hence the contribution of electromagnetism to the nuclear mass, once the masses of the individual neutrons and protons are known. Similarly, a complete theory of hadron structure would enable the electromagnetic (and weak) mass splittings of the individual hadrons to be computed.

In section 6.3 we shall see that recent evidence indicates how hadrons do have a complicated internal structure, but little progress has been made in a quantitative description. All that we can say is that the observed magnitude of the mass splitting is of the right order for the force involved. For example, the mass difference between π^0 and $\pi^{\pm}$ is 4.6 MeV, or about 3% of the rest mass. The percentage splittings for the kaons, Ξs and sigmas are 0.8, 0.5 and 0.7, respectively. These numbers are all in the region of 1/137, as expected. The weak splitting of 10^{-11} MeV for the K_1^0, K_2^0 is around 10^{-14} of their rest mass, and is typical of the coupling strength of the weak interaction (see Section 6.4).

All this raises the fascinating prospect that other symmetries of nature may lie buried among the subatomic particle data, but are better concealed by more powerful symmetry breaking than even electromagnetism. Suppose, for example, that part of the *strong* interaction is responsible for breaking a hitherto undiscovered, hidden symmetry. Crude reasoning suggests grouping still more particles into multiplets containing mass differences some 100 times larger than even the electromagnetic splitting. Thus we might think of grouping together particles with masses that differ by as much as a few hundred MeV, and regarding them all as simply

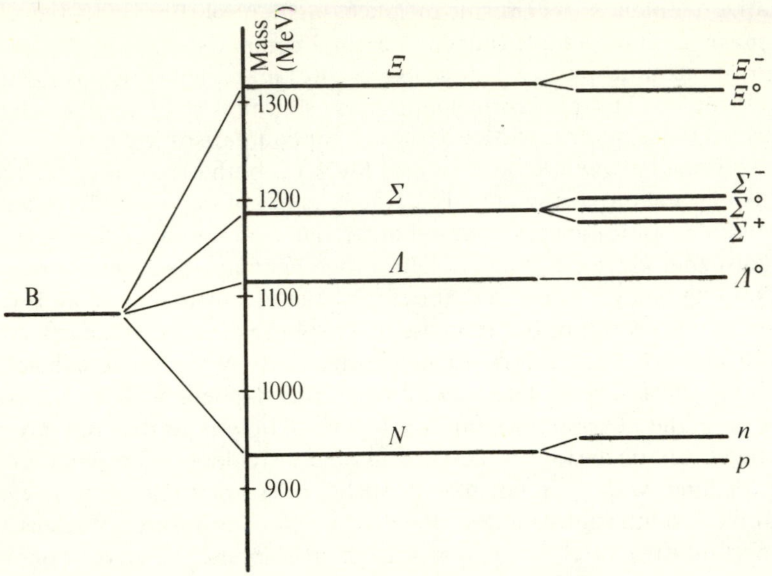

Fig. 6.2. These eight baryons are considered as split states of a single particle B. A hypothetical semi-strong force first splits B into four components, Ξ, Σ, Λ and N, several hundred MeV apart. The feebler electromagnetic force then splits these into charge multiplets with masses a few MeV apart. Similar effects due to weak or gravitational forces, such as that which causes the minute K_1^0, K_2^0 splitting, are swamped. Notice that the higher mass particles have the greater strangeness.

different states of a single particle, labouring under a strong force which acts unequally on the various states.

Consider, for example, the eight baryons n, p, Σ^+, Σ^0, Σ^-, Λ, Ξ^0 and Ξ^-. All have spin one half and baryon number one, though they differ by strangeness. All lie within 400 MeV of each other. Fig. 6.2 shows how a single baryon B can be considered as split by a semi-strong and an electromagnetic interaction into successively finer multiplets. A significant feature, which we shall return to later, is that the higher mass particles have the greater strangeness. Evidently the part of the strong interaction which splits the baryons can see strangeness. The rest of the strong interaction is strangeness-blind. A similar grouping can be made of all the hadronic mesons.

In 1960 Murray Gell-Mann and the Israeli physicist Yuval Ne'eman showed how the eight baryons could be arranged into a suggestive pattern, known as the eightfold way (after Buddha), shown in Fig. 6.3. The rows are labelled by strangeness and the columns by isotopic spin projection (i.e. the component of the spin along the 'up–down' axis). Thus, the top row is the isotopic spin doublet n,p, the middle row

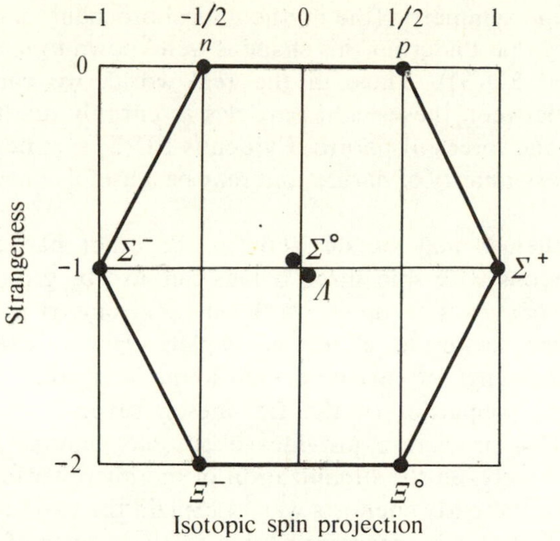

Fig. 6.3. The baryon octet of Fig. 6.2 is here arranged in a fashion which suggests an underlying symmetry principle at work. We plot strangeness against isotopic spin projection.

contains the triplet $\Sigma^+, \Sigma^0, \Sigma^-$ and the bottom row Ξ^- and Ξ^0. The Λ has isotopic spin zero, so stands as a singlet at the centre of the pattern, with the Σ^0.

The top row has strangeness zero, the second row minus one, the third row minus two. The small horizontal mass splittings are presumably caused by the electromagnetic force, and the larger vertical splittings by the semi-strong part of the strong force, which discriminates between states of different strangeness. Notice that if one looks along the oblique lines one sees that they connect states with the same electric charge.

It is a curious thought that according to this scheme, the difference between the protons and neutrons, out of which the solid and enduring atoms of the world are built, and the hyperons, which disintegrate in about 10^{-10} s, is only one of mass. If strangeness had caused a *lowering* of mass levels rather than raising them, then the protons and neutrons would be the transitory creatures, and the world would be built out of Ξ hyperons instead.

The hexagonal figure formed by this family is reminiscent of the shapes drawn in Fig. 5.2 and discussed in the section on symmetry. Indeed, the symmetry of Fig. 6.3 is apparent at a glance. If the weak, electromagnetic, and strangeness-discriminating part of the strong interaction were switched off, we might expect all the particles in the figure to have the same mass. We could then say that the laws of physics were 'hexagon' symmetric (the mathematical procedure associated with the symmetry that underlies this shape is well known to mathematicians and is called SU(3)). Thus, in the real world, we can regard the differences between these eight particles as entirely due to *symmetry-breaking* by the forces of nature. Evidently SU(3) symmetry is only an approximate symmetry of nature, but may be a useful concept, nonetheless.

Similar schemes may be devised for the other particles, e.g. the mesons. Baryons with spin three halves can also be grouped together into a family of *ten* – a decuplet – with similar symmetry properties. Still others appear to reside alone as singlets. By an examination of symmetry groupings of this type Gell-Mann was able to predict the existence and properties of the Ω^- meson several years before its discovery. This prediction provides a graphic demonstration of the power of symmetry in the organization of the microworld.

The goal of the early chemists was to explain the varied and complex properties of matter by searching for a small number of fundamental particles. Out of the union of these supposedly indestructible atoms of matter, all the rich diversity and variety of ordinary objects was to be accomplished. This century, physicists have smashed the atom, and had to begin a new search at a deeper level for truly *elementary* building blocks of matter. With the alarming proliferation of species in the subatomic zoo, this search seems hopeless. With every improvement in technology, more and more particles turn up to confuse us, and the array of available species threatens to become totally bewildering. Can there remain any hope of building all these particles out of a few elementary ones? The particle multiplets described in this section suggest a deep symmetry and simplicity at the heart of all matter. We seem to be glimpsing something important – the suspicion of some really fundamental units somewhere below the surface muddle of the data.

Mankind has long been fascinated by numerology. The bible frequently refers to the numbers 7 and 40, which appear to have held some sort of mystical significance for early Judaism. In Buddhism, there are 8 commandments to relieve mankind of all pain. Christianity is founded on the concept of a trinity, and the work of 12 apostles. In fundamental science too, strange numbers occur, some of which have been described in Section 2.8. The particle multiplets throw up the numbers 10, 8 and 1. What is the reason for this? How can this numerology be understood and the underlying symmetries analysed and sharpened?

6.3 *Quarks: charmed, coloured, strange and flavoured*

The study of symmetry has a long tradition among mathematicians. Particularly important was the work of the nineteenth-century Norwegian Sophus Lie (1842–99). It was natural that, a century later, physicists should turn to his ideas for guidance. One attractive feature of this work is that there exists a mathematical process which does for symmetry arrangments what multiplication does for numbers. Just as multiplying small numbers gives larger ones so multiplying simple symmetries gives complicated symmetries. This idea is already familiar from the study of crystal lattices. The regular shapes of crystals can be explained as a *combinatorial* effect of billions of smaller, symmetric atomic units.

Just as groups of symmetric atoms build up crystals with interesting symmetries, so too perhaps there is some 'atomic' or elementary structure out of which the baryon and meson singlets, octets and decuplets can be built? This idea was pursued independently by Gell-Mann and the American physicist George Zweig, who in 1963 proposed that all hadrons were ultimately composed of unions of just *three* truly fundamental particles, which associate together in simple symmetry groups. Gell-Mann coined the name quarks (after a line in James Joyce's novel Finnegan's Wake) and it has stuck. Leptons are not composed of quarks; in this scheme they are considered to be 'already elementary' in their own right. Sometimes it is helpful to regard leptons and quarks as point-like entities, as they are not supposed to have any internal structure at all.

In the world of subatomic particles, all particles of the same species are completely identical and indistinguishable. The only properties that label a particle are those features that define the species: spin, electric charge, baryon number, lepton number, strangeness, and so on. These numbers only take certain discrete (usually integer) values, so they are usually called quantum numbers. In the quark theory, an attempt is made to reproduce the known quantum numbers of all the hadron species out of combinations of quarks. The quantum numbers of the assembled hadron are obtained simply by adding those of the constituent quarks (with the exception of spin, which, being a vector, must be combined using more elaborate rules).

The essential features of quark combinatorics are easy to grasp. Quarks are fermions which combine together in pairs or triples. Mesons, which are bosons, are made of quark–antiquark pairs, giving a total spin of zero or one unit. Baryons, which are fermions, are composed of three quarks (antibaryons of three antiquarks), so they have a combined spin of one half or three halves.

The choice of three for the number of quarks is a natural outcome of the study of the way in which symmetry groups are multiplied. For example, since a meson is made up of a quark and an antiquark and there are three different types of quark, there will be nine (3×3) possible combinations that can make a meson. When symmetry groups are multiplied, their larger groupings often reduce to sums of smaller ones. Thus, $3 \times 3 = 9 = 8 + 1$; it turns out that the nine combinations reduce to a family of eight (an octet) and a solitary meson (singlet). Similarly, the baryons may be formed out of 27 ($= 3 \times 3 \times 3$) combinations, which reduces to $1 + 8 + 8 + 10$, which accounts for a singlet, two octets and a decuplet. These reduced combinations of groups therefore correspond precisely to the favoured family numbers which are observed.

Clearly it is only possible to produce an assembly with a single unit of electric charge by endowing the quarks with *fractional* charge of 1/3 or 2/3, a bizarre idea that would instantly identify a solitary quark should one be encountered. To make baryons, each quark must carry a baryon number of 1/3 (each antiquark has $-1/3$). Also, to account for strange particles, one quark must carry strangeness of one unit.

The quarks are given the arbitrary but simple names 'up', 'down' and 'sideways', denoted by u, d and s, respectively. These labels are called the quark *flavours*. The u quark has a charge 2/3, while d and s each have $-1/3$. The s quark carries -1 strangeness. Naturally the three antiquarks have the reverse charge and strangeness. The arrangements of quarks for any particular hadron are easily guessed by specifying the spin, charge, baryon number and strangeness. For example, the proton is composed of two u quarks and a d, giving spin 1/2, charge $+1$, baryon number $+1$ and strangeness zero. Proceeding in this way, the whole octet of baryons may be constructed: symbolically, $n = udd$, $\Sigma^+ = uus$, $\Sigma^0 = uds$, $\Sigma^- = dds$, $\Xi^0 = uss$, $\Xi^- = dss$, $\Lambda = uds$. The mesons can be composed of quark-antiquark pairs. For example, $\pi^+ = u\tilde{d}$, $K^+ = u\bar{s}$ and so on. Proceeding in this way all hadrons known before 1974 could be incorporated in the scheme. What is more, every combination corresponded to a known hadron – there were no gaps (see Fig. 6.4).

By supposing that the strange quark is considerably heavier than the u and d quarks, it is possible to give a simple explanation of why the hadrons with a high strangeness are generally more massive. Consider, for example, the heaviest spin one half baryon, the Ω hyperon. With a mass of 1672 MeV it is more than 300 MeV heavier than the next lightest hyperon, Ξ^-, which is in turn 124 MeV above the Σ^-. According to the quark scheme Σ, Ξ and Ω contain one, two and three s quarks, respectively.

Quarks are supposed to be bound together by the strong interaction in some as yet mysterious way. The combinations will have quantum

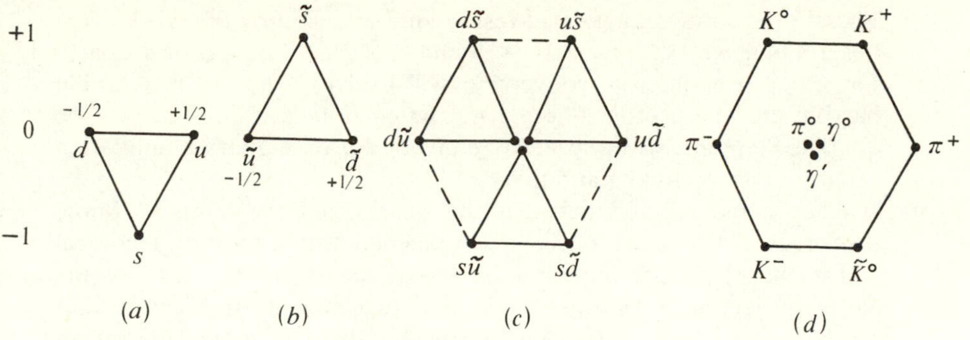

Fig. 6.4. Quark symmetries. (*a*) The triplet of quarks *u*, *d* and *s* can be arranged in a triangle according to strangeness and isotopic spin, in a fashion similar to the arrangement in Fig. 6.3, but simpler and more elementary. (*b*) The three antiquarks form a reflected triangle. (*c*) Multiplication of the simple triangular symmetry groups (*a*) and (*b*) leads directly to the more elaborate hexagon of the type shown in Fig. 6.3. (*d*) This nonet (8 + 1) of mesons corresponds exactly to the quark combinations shown in (*c*). Thus the strange particle K^+ has quark content $u\tilde{s}$ giving a charge of 2/3 + 1/3 = +1 and strangeness +1. A further group multiplication reproduces the baryon hexagon of Fig. 6.3.

energy levels similar to those of atoms, and excited states may be produced which rapidly decay. These are the resonances. In atoms the energy levels are spaced out by a few electron volts; here the levels are hundreds of MeV apart!

It might be wondered how a pion, composed of supposedly elementary quarks, can decay into a muon and a neutrino, which are both leptons. Where have the quarks gone? The explanation is that the weak interaction can *change the flavour* of quarks. Thus π^+, which consists of the quark pair $u\tilde{d}$, can be converted to $d\tilde{d}$ or $u\tilde{u}$, which annihilate, quark against antiquark. The released energy then goes into creating the leptons (or, more completely, a *W* boson is first created which then decays into a muon and neutrino). The decay of the neutron is another example of a change of flavour, from *d* to *u*, under the weak force. Flavour change brought about by the weak force also explains how strange particles decay into non-strange progeny. The *s* quark emits a *W* and converts into a *u* or *d* quark (see Fig. 6.5).

In contrast, quark flavours are left unchanged by the strong interaction which is, after all, responsible for binding them together in some way. During strong transmutations, the *s* quarks can be made in $s\tilde{s}$ pairs, because the total strangeness is zero. The pair can then be split, with $\tilde{s}$ going to make up a particle with strangeness +1 and *s* going to make up another particle with strangeness −1. During the subsequent decays of the strange particles, the *s*, or $\tilde{s}$, quark must be passed on from one particle to another under any strong interaction transmutation until the lightest strange particles – the *K* mesons and the Λ baryon – are formed.

Therein the strange quark will reside until attacked by the weak force (after a long wait of maybe 10^{-8} s) and converted into a u or d quark. For the Λ, this internal conversion will lead to a non-strange stable baryon, e.g. the proton. For the K, the essential end-product will be quarkless leptons, with the newly converted u or d quark annihilating against a corresponding antiquark.

What is the relation between the quarks and the leptons? Many theorists feel that there must be a fundamental link on both technical and aesthetic grounds, but the imbalance of having three quark flavours and *four* leptons (electron, muon and two neutrinos) suggests that maybe a quark flavour has been overlooked. Is there room for a fourth quark? In 1970, arguments were advanced that certain unexpected regularities in weak decays required the participation of an additional quark. It would form a natural companion to the strange quark, in the same way that the up and down quarks form a natural pair. It would have to be distinguished from the other three by a new flavour, which is some what whimsically known as *charm*. The charmed quark is denoted c, and has charm $+1$ and charge 2/3. Charm is conserved in all strong and electromagnetic processes, but not weak. The antiquark $\bar{c}$ has charm -1. The existence of charm was first suggested in 1964 by the American physicists Sheldon Glashow, James Bjorken and others.

Evidence for charm came unexpectedly in 1974. Two groups of workers, one headed by Samuel Chao Chung Ting at MIT and the Brookhaven National Laboratory, and the other by Burton Richter of the Stanford Linear Accelerator Center and the Lawrence Berkeley Laboratory, independently discovered a remarkable new particle. It is a spin one, non-strange, electrically neutral meson, with a huge mass more than three times that of the proton (3097 MeV). It decays into pions, muons, electrons and other familiar particles. What is so odd about the new particle is its lifetime, which for a non-strange hadron is some 1000 times too long (about 10^{-20} s). Known as the psi (ψ) particle, its curious longevity means that it shows up in collision experiments (e.g. electron–positron annihilation) as a very narrow resonance in the scattering cross-section – so narrow that it had been overlooked before. In 1976 Ting and Richter received a Nobel prize for their discovery.

The theorists were ready with an explanation. The three-quark model, so successful in accounting for the then known hadrons, simply had no room for the ψ. A fourth, still heavier, quark was needed – the charmed one. Evidently ψ consists of a charmed quark–antiquark pair, $c\bar{c}$. Moreover, the longevity of $c\bar{c}$ has a precedent in the so-called phi (φ) meson, made up of a strange quark–antiquark pair $s\bar{s}$. The φ has a lifetime about 10 times longer than usual for a strongly decaying hadron.

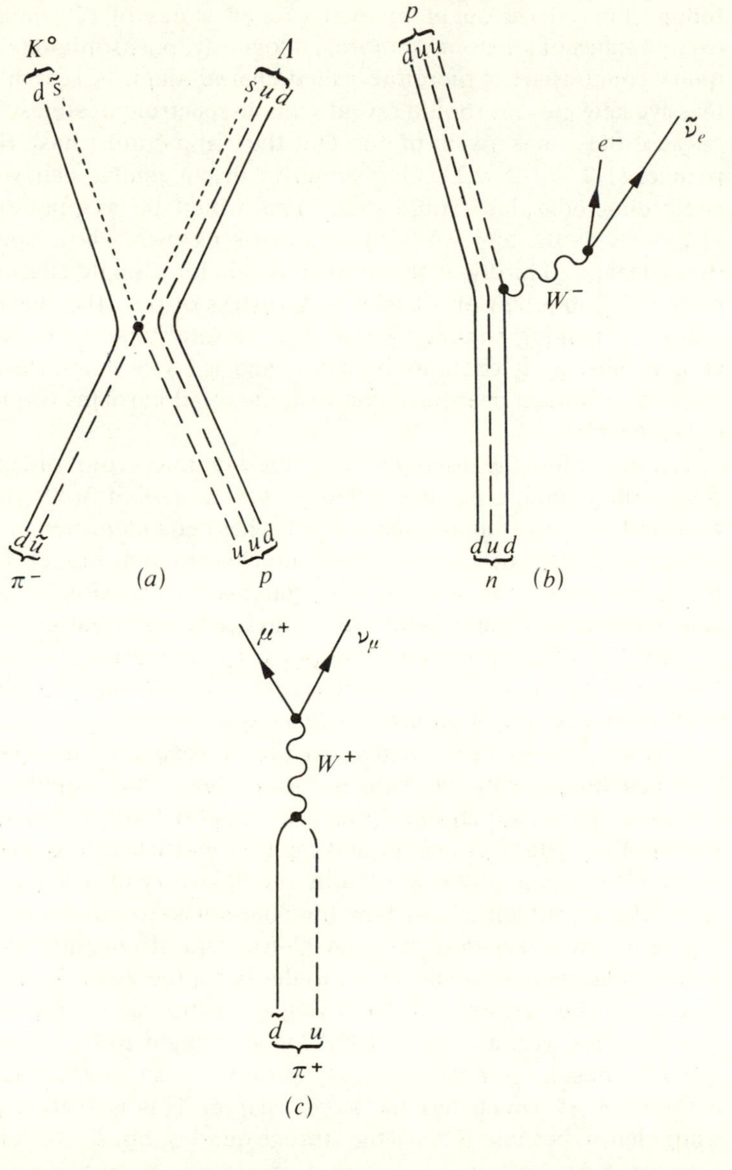

Fig. 6.5. Formation and decay of particles. These processes have a natural explanation in terms of quarks. (*a*) $p + \pi^- \rightarrow K^0 + \Lambda$. A u quark from the proton annihilates a $\tilde{u}$ antiquark from the pion and creates an $s\tilde{s}$ pair, which divides between the two strange particles K^0 and Λ. (*b*) Decay of the neutron. The emission of a W^- changes a d to a u quark (i.e. $n \rightarrow p$), and decays into leptons. (*c*) The pion decays weakly into leptons via a W^+. The weak interaction converts a quark to the flavour of the companion antiquark, which enables the quark content to annihilate completely.

Assuming this explanation is correct, several interesting predictions follow. First there ought to exist excited states of $c\bar{c}$ similar to the excited states of an atom, or more analogously, positronium (e^+e^-). The quark counterpart is therefore called charmonium. A search for more massive new mesons should reveal a whole spectrum of states. Secondly, because the ψ has a spin of one unit the c and $\bar{c}$ must have their spins parallel (1/2 + 1/2 = 1). One supposes that a similar pair with oppositely directed spins should exist. This would be a spin zero meson (1/2 − 1/2 = 0), and would also possess its own whole spectrum of excited states. Thirdly, it should be possible for a single charmed quark to combine in a pair or a triplet with quarks of the other flavours, u, d and s, so making *charmed particles*. The total charm of the ψ is, of course, zero as it contains both a c and a $\bar{c}$, so ψ is referred to as containing hidden charm. In contrast, charmed particles would contain *naked* charm.

Ten days after the discovery of ψ, the Stanford group discovered the first excited state, a particle called ψ' with a mass of 3685 MeV. Since then a whole spectrum of other particles has been identified, in beautiful accordance with the predictions of the charmonium model. For example, a few ψ''s are found to decay by gamma ray emission to form a spin zero particle at around 3400 MeV. Physicists investigating the gamma ray spectra of charmonium are repeating at ultra-high energies the history of the atomic spectroscopists who studied the photons of light from excited states of atoms 100 years ago.

Following these spectacular successes, experimenters frantically searched for nakedly charmed particles, those that would contain an unpaired charmed quark or antiquark. Six possibilities for new mesons present themselves: $c\bar{u}$, $c\bar{d}$, $c\bar{s}$ and their antiparticles. A search for such mesons was eventually rewarded by the discovery of new particles, but there was a problem about how the charm was to be identified. If the suggested close relation between charm and strangeness is properly understood, then a c quark ought to decay via the weak interaction into an s quark, but *not* an $\bar{s}$. Good evidence for this comes from one of the new particles with a mass of 1868 MeV, thought to be $c\bar{d}$, now called the D^+ meson. It decays weakly into $K^- + \pi^+ + \pi^+$, but not into $K^+ + \pi^+ + \pi^-$ which has the same charge. This is readily explained using charm, because K^- has the strange quark s, but K^+ the antiquark $\bar{s}$ (see Fig. 6.6).

The extraordinary success of the quark model in tidying up the proliferation of hadrons is marred by only one shortcoming, and it is devastating. Nobody has ever seen a quark. No experiment, however energetic, has been able to break a hadron into its constituent quarks. In spite of years of search and several false alarms, no quarks have been

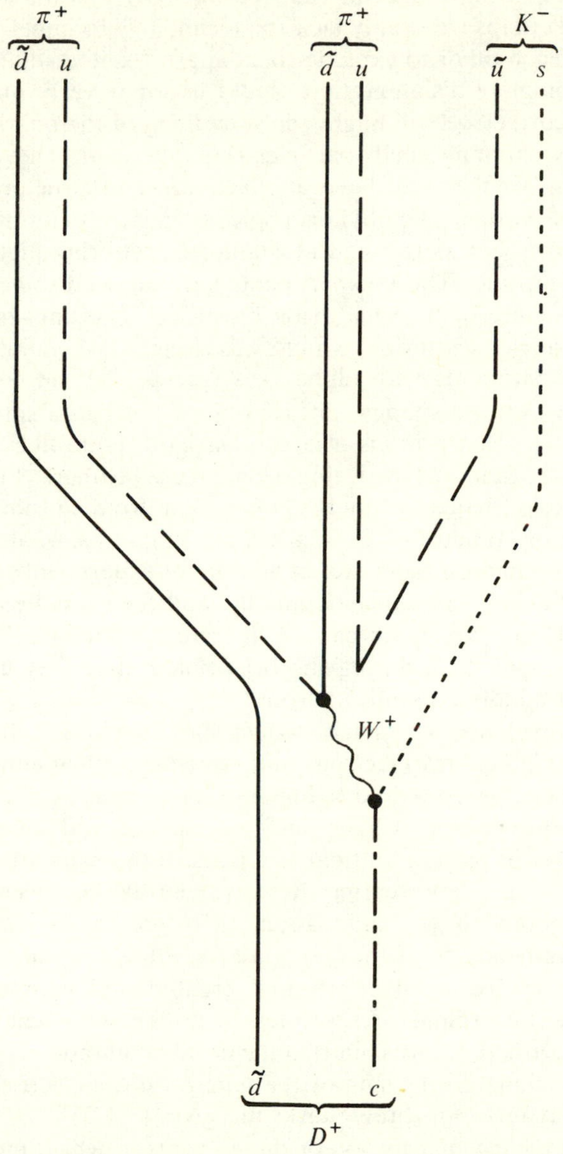

Fig. 6.6. Decay of a charmed particle. The weak force converts the charmed quark c into a strange quark s, which becomes part of the strange meson K^-. The W^+ boson which achieves this transmutation then breaks into a $u\tilde{d}$ pair which divides among two pions.

detected. Do they really exist?

Some physicists have argued that splitting the hadron might be as hard for us as splitting the atom (i.e. the nucleus) was for our great grandparents. Perhaps we simply lack the technology to impel particles fast enough at each other to explode them apart. Excited states can be created, but complete disintegration eludes us for poverty of energy. Even if this is correct we still might see something of the quarks *inside* the hadrons by shooting small particles *through* them. The particles must be leptons, or they will have as much size and structure as the target we wish to probe. The obvious choice is electrons and important experiments have been performed at Stanford, scattering high-energy electrons from protons. These experiments have enabled something of the interior structure of hadrons to be discerned. The unusually high level of wide-angle scattering which was found is reminiscent of Rutherford's experiments with alpha rays (see p. 44); he concluded that the atom was not a spongy, diffuse object, but had a small, hard nucleus. It appears that protons also contain hard, point-like particles responsible for the Stanford scattering. Could these be quarks? Perhaps.

Other indirect evidence for quarks could come from considering the following question. If hadrons are really made of quarks, what happens when hadron–antihadron pairs are created in an experiment? Presumably quark–antiquark pairs appear initially, and then rapidly combine into hadrons. Even though we cannot liberate the individual quarks, perhaps we can spot them during the brief phase after they are born, before they get bundled up into hadrons?

This sort of evidence for quarks would show up in a study of the annihilation of high-energy electrons and positrons. When annihilation occurs, all the energy goes into a high-energy gamma ray. It is very probable that this photon will be only virtual and will immediately produce a shower of particle–antiparticle pairs. If the pairs are leptons, for example $\mu^+ + \mu^-$, then treating these as point-like particles enables accurate predictions to be made about their production rate using quantum electrodynamics. If the pairs are hadrons, we could likewise assume that point-like quarks are first created and compute *their* production rates accordingly. Experiment will then show whether this assumption is justified. It turns out that the production ratio of quarks to muons is proportional to the sum of the *squares* of the electric charges on the quark flavours. For three quarks this gives $(-1/3)^2 + (-1/3)^2 + (2/3)^2 = 2/3$, implying that for every three events in which muons are produced, there should be two in which hadrons (quarks) are produced.

Experiments in the early 1970s gave a ratio of 2 rather than 2/3, and for this reason and others theorists began to question whether they had counted up the quarks correctly. Even allowing for the fourth, charmed, quark, the discrepancy is large. There was also a theoretical reason for doubt. A particle such as Ω^- contains three identical quarks (*sss*) with spins aligned (total 3/2). However, quarks are fermions, so the Pauli exclusion principle should forbid three such identical particles to occupy the same quantum state. To avoid these problems, it was suggested in 1964 by the American physicist Oscar Greenberg that each flavour of quark actually comes in three different varieties, arbitrarily called *colours* (no connection with ordinary colour is implied). The names chosen are red, green and blue (R, G, B). At first sight the increase in the number of quarks from four to twelve seems to imply a vast and unwanted increase in the total number of possible quark combinations, but this is not so. Unlike charm, which can appear in an assembled particle, colour is strictly hidden from view by requiring the assembled hadrons to be always 'colourless'. To achieve this, baryons are composed of one quark of each of the three colours: red + green + blue = white. In this way the Pauli principle is satisfied because the quarks are no longer identical particles – they are distinguished by their colours. Colourless combinations can also be made by supposing that an antiquark carries 'anticolour'. Thus, the mesons contain one coloured quark and a corresponding antiquark of the appropriate anticolour. The colour does not matter: one should think of the pair as continually and simultaneously changing between red–antired, green–antigreen and blue–antiblue, in a sort of multichrome superposition.

Because all assembled hadrons are colourless, the introduction of this purely internal property does not lead to any increase in the number of permissible combinations: evidently nature is colour-blind. However, it does increase the number of distinct quarks from four to twelve. Consequently, when calculating the hadron–muon production ratio described above, we must multiply by three. Now the charmed particles are all extremely massive, so evidently the charmed quark is considerably heavier than the others. It follows that at low energies charmed hadrons will not be created, so the hadron–muon ratio should be $2/3 \times 3 = 2$, precisely as observed. At higher energies, when charm begins to appear, the ratio should start to rise, and this is also observed in the experiments. The charge of *c* is 2/3, so three coloured, charmed quarks contribute an extra $3 \times (2/3)^2 = 1/3$ to the ratio, giving 3 1/3 in all. In fact, the experiments show a rise to about 5 above 4000 MeV, so it appears that still more particles may be required. We shall return to this subject in a later section.

Right from the outset, physicists have always assumed that if matter is composed of more elementary constitutents, then it can eventually be broken apart. All that is needed is enough energy (which means enough money to build a sufficiently large accelerator). Recently, though, some

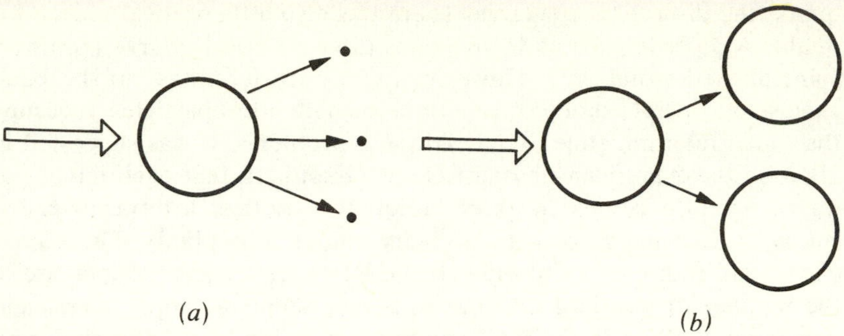

Fig. 6.7. (*a*) When a hadron is smashed into bits and pieces by a high-energy particle we might expect to see quarks spill out among the debris. (*b*) Instead we see only more hadrons.

theorists have begun to wonder whether this is necessarily so (see Fig. 6.7). Could it be that quarks are a reality, but that they can *never* be removed from hadrons, however much energy is available? The idea of permanently confined quarks does not stem solely from the continuing inability of the experimenters to free them. Many of the properties of hadrons seem to suggest that the quarks inside them behave at high energies as though they are rather *loosely* bound.

How do loosely bound quarks avoid getting knocked out of hadrons under the ferocious impacts of the high-energy electron scattering mentioned above? Return for a moment to the concept of the electric force field which binds an electron to an atom. At close distances to the nucleus the electron is tightly bound, but it can be freed with sufficient energy because the electric force falls off with separation (as $1/r^2$). The electron gets looser and looser as it is drawn away from the nucleus. Contrast this diminishing grip with that of a piece of elastic. Two particles bound by elastic cannot be pulled asunder because the elastic tension *rises* as it is stretched. The particles are bound by a force which actually increases with distance. But when they are close together the force is small and the particles are loosely bound. Of course, real elastic eventually snaps, but quarks could be bound together by a super-elastic force that never 'snaps'. They would then be trapped for ever.

In addition to this, quark confinement has a neat description in terms of colour. If nature refuses to display the true colour of quarks, then a solitary coloured quark cannot be seen. Nor, for that matter, can a combination of say, four or seven quarks. Only a red–green–blue triplet (or an oppositely colour-matched quark–antiquark pair) can unite in a colourless way. This also explains why *leptons* are found singly and not united in doublets or triplets – they are individually colourless, so can exist in isolation.

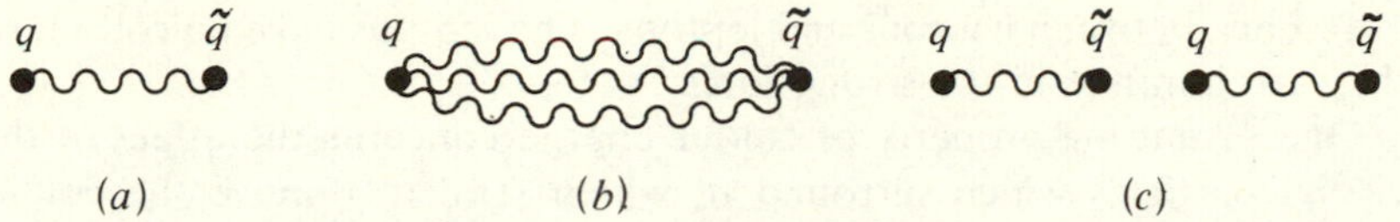

Fig. 6.8. Pulling hadrons asunder. As the meson is stretched the binding force between the quarks (q) grows, until the bonds have stored enough energy to supply the mass of a new quark–antiquark pair. The meson then breaks into two mesons, *not* two quarks. The situation is similar to that shown in Fig. 1.8, where attempts to pull magnetic poles out of bar magnets leads to two magnets, *not* two poles.

Fig. 6.8 shows what happens as the two quarks in a meson are pulled farther apart. The binding force rises, and the energy stored in the bonds increases, until a point is reached when it is energetically more favourable for a new quark–antiquark pair to be created than for the stretching to continue. The result is that in a sense the meson snaps, but instead of breaking into two isolated quarks, the two new quarks combine with the old to form *two* distinct mesons. Clearly, however much energy is available, it would never be possible (according to this model) to pull away a single quark, so it is no surprise that they are never observed.

What is the nature of the force that binds the quarks together? At this stage there is much conjecture but little rigorous theory. An obvious approach is to suppose that there is a new field of force, with the quarks carrying a new type of charge and interacting through the exchange of new types of quanta. The latter have been dubbed 'gluons' because they provide the hadron glue, just as Yukawa envisaged the mesons providing the nuclear glue. The strong force *between* hadrons is regarded as but a feeble vestige of this powerful *internal* force.

In one model the new type of charge is in fact, colour, and the gluons are massless, spin-one particles like photons. The situation is thus reminiscent of quantum electrodynamics, and has become known as quantum chromodynamics. However, chromodynamics is complicated by having *three* colour 'charges' as opposed to one type of electric charge. The colour fields associated with this multiplicity are interwoven in a complicated way which demands no less than *eight* types of gluons to communicate the strong force. A mathematical study shows that in such a multicomponent system the gluons must also be coloured (contrast the photon which carries no electric charge), so the quarks change colour (though not flavour) when the gluons are emitted or absorbed. Because of the continual exchange of gluons the quarks do not remain of one colour, but pass rapidly through red, green and blue. It is not possible to say which quark is which colour at any given time. If the strong force is really due to colour, it elegantly explains the difference between hadrons and leptons. The leptons have no colour, so they are immune to the strong force.

One promising property of colour charge concerns the effect of the virtual particles which surround it, which tend to remove the colour from the quark itself and spread it out in a cloud. Thus, when two quarks are very close together they do not feel much force as they are 'inside' each other's colour charge. So at small separations which are probed using high-energy projectiles they behave like a collection of free particles, precisely as observed. At greater separations the charge is noticed and the force rises steadily, binding the quarks together more firmly (though it is not yet known whether it is firm enough to provide permanent confinement). Thus, under low-energy bombardment the hadron behaves like a cohesive integral particle (see Fig. 6.9).

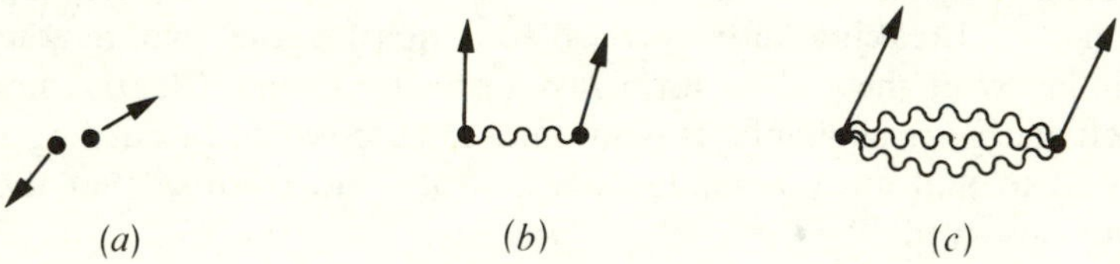

Fig. 6.9. (*a*) At close distance the quark binding disappears and the particles move independently. This is the situation observed under high-energy electron scattering from hadrons, which probes the short-distance behaviour. (*b*) At intermediate distances the gluon force binds the quarks loosely. (*c*) At large separations the glue sticks the hadron together into a rigid body. In this stiffened condition, which is examined under *low*-energy bombardment, the hadron moves as an integrated unit.

In another theoretical scheme, the force between quarks is modelled more closely on strings of elastic (the 'string model'). Yet another envisages the quarks as bouncing around in a sort of elastic bag, like molecules in a balloon.

Although the success of the quark hypothesis is impressive, and a large number of hadron properties are explained by it, the limitations are obvious. What started as a simple collection of three particles has grown and grown. The quarks come in different colours, with and without charm, there are probably several types of gluons needed to bind the quarks together, and permanent confinement means that neither quarks nor gluons can ever be observed individually, even in principle. Moreover, the nature of the confinement is not really understood at all, the proposed ad hoc models appearing very rudimentary and contrived. To make matters worse, evidence seems to be mounting for the inclusion of a fifth, perhaps a sixth quark. Where will it end? Does not such a proliferation of quarks and gluons make nonsense of the claim that they are 'elementary' particles? Shouldn't we look for some other subunit out of which to build the quarks? Does this uncovering of layer upon layer of structure ever end, or does the hierarchy of matter continue *ad infinitum*?

These questions cannot be answered, but we can explore alternative perspectives. One of these is the idea of particle *democracy* proposed in 1961 by the American physicists Geoffrey Chew and Steven Frautshi. To understand the concept it is helpful to review what is really meant when it is said that a particle is 'made up' of others. In earlier sections we saw how no particle remains inert. The probabilistic nature of quantum mechanics obliges all particles to engage in a restless dance of activity, emitting and absorbing virtual quanta that are *themselves* particles. For a stable particle like a proton or an electron, the nature of the central particle itself seems somehow more substantial than the evanescent cloud of virtual particles which surround it, but for hadrons that live only 10^{-23} s there is little distinction between real and virtual.

Moreover, we have seen that all particles are apt to flip their identities from time to time, a photon becoming an electron–positron pair, a pion a proton–antiproton pair, and so on. In this shifting world of transmutation, is it meaningful to pin down a particle and say that it is such-and-such an entity? If we were to consider progressively higher perturbations – Feynman diagrams with ever more lines and vertices – we should be forced to admit that every particle has mixed into it somewhere something of every other type of particle, however dilute.

In the macroscopic world we have no difficulty in separating the identities of things: a cat is a cat, a dog a dog; neither is a horse, or a star or a rock. In the quantum domain, identity becomes less definitive. When we talk of a particle being 'made up' of others we do not really mean they are lumped together in the macroscopic sense. Quantum particles are described by *undulations*, so that particles combine with others after the fashion of waves, with their ripples superimposed in a complicated interference pattern. One can talk only of the probability that a measurement would detect a photon, or an electron–positron pair, or a $\pi^+ \pi^-$ pair, etc. *All* these eventualities are potentially realizable with a certain likelihood, some more than others, but all possible.

Viewing the microworld this way it seems conceivable that there are actually no such things as 'elementary particles' at all. No one particle is more fundamental than any other. Every particle is made up of every other. All particles are equal. This is the subatomic democracy. Can it work as a real theory, or is it just an aid to visualization?

Geoffrey Chew and Steven Frautschi examined the idea in detail. To illustrate the essential feature recall the nature of the strong force inside the nucleus. Yukawa first explained it as an exchange of virtual pions. We now know that pions are only one of a whole spectrum of mesons, and more massive ones, such as the ρ meson, are also responsible for part of the nuclear force. Their greater mass means their range is shorter, so they contribute to the short-distance behaviour of the nuclear interaction.

Now *all* hadrons feel the strong force, so must likewise exchange virtual mesons. This remains true of the *mesons themselves*. The force between two pions occurs from, among other things, the exchange of virtual ρ mesons. Sometimes the pion–pion force is strong enough to bind them together briefly into a composite di-pion. In the spirit of subatomic physics this too is a particle. What particle? The ρ meson! In Section 6.1 it was pointed out that ρ is really a union of two pions. So we see that the exchange of virtual ρs helps bring about the formation of a ρ.

There is, of course, no beginning or end to this chain. Pions are made of ρ mesons which are made of pions which are . . . If the theory were perfect, it would prove that there is only *one* way in which the infinity of bits of the jig-saw will fit together, and so reveal that only with the exact masses and force strengths that we observe is the whole scheme self-consistent. We should then have a complete explanation of matter, with all the particle properties fixed by the internal consistency requirement. Unfortunately there is more to the world than π and ρ, and a detailed theory of *all* the interacting hadrons is way beyond present hopes. Nevertheless, the concept is novel and reminds one of the story in which a man hauls himself out of a bog by pulling on his own bootstraps. For that reason the Chew–Frautschi idea is called 'bootstrapping'.

6.4 *Unified forces*

Turn once again to the list on p. 4. At the time of Newton the world appeared to exhibit hundreds of different kinds of forces. Today, they have been unified into only four. Mathematically, the most significant unification was that of electricity and magnetism into Maxwell's laws of electromagnetism. The question naturally arises as to whether a further reduction in number is possible by combining the remaining four together in some way. The ultimate achievement would be a single mathematical theory encompassing all the forces of nature.

The forces do appear to share a number of common features. They are all mediated by the exchange of bosons: the graviton, which has spin two and is massless, the W boson, which has spin one and is very massive, the photon which has spin one and is massless, and the gluons which also have spin one and are massless.

The activities of the strong force are hemmed in by very many conservation laws. Besides the usual classical laws, such as energy conservation, the hadronic interaction conserves strangeness, isotopic spin, charm, mirror symmetry (parity P), time-reversal symmetry, charge reflection symmetry (C), baryon and lepton number (both kinds) and maybe others which we do not know about. The electromagnetic force is almost as well behaved, though it violates isotopic spin conservation. The weak force is the maverick. It seems to violate every law going: strangeness, charm, parity and time reversal, not to mention isotopic spin. It seems, however, to respect baryon and lepton number. In this scheme, there is a tendency for the weaker forces to have greater freedom of action, so perhaps gravity violates even these latter sacrosanct laws? It is too weak for us to experiment with, but the theory of black holes does suggest that most of these laws are violated or at least transcended as far as these enigmatic objects are concerned.

Consider first the force that has the most successful theoretical basis: electromagnetism. Starting with the work of Planck and culminating with Dirac, Feynman and others, quantum theory was applied to the electromagnetic field and to its interaction with charged particles. Quantum electrodynamics is the archetypal relativistic quantum field theory. The range of results and their astounding accuracy are without precedence in physical science. It is by far the most successful physical theory of any sort in existence.

Coming to the weak interaction, we have an attractive theory that has all the ingredients of success and which accounts for a wide range of low-energy processes. Some mysteries, such as the way in which the field couples to hadrons, remain. However, there are deep theoretical reasons why the theory of weak interactions as outlined so far cannot be correct. These reasons will be explained below. For now we simply note that it is the weakness of the force which enables sensible predictions at low energies to be made by treating the force as a small perturbation, and that if we try to calculate higher-order corrections we get ambiguous nonsense.

Gravity has a most beautiful and comprehensive classical description due to Einstein, in terms of the geometry of space–time. This special type of description already sets gravity apart from the other forces of nature. Moreover, in spite of its healthy classical shape there is no successful theory of quantum gravity. Strenuous efforts over the past two decades have failed to produce a quantum theory of gravity based on small perturbations which gives meaningful answers above the lowest orders, a shortcoming which it shares with the Fermi theory of the weak interaction.

Finally, our theory of the strong interaction is in a very rudimentary state. There is no classical limit, and no real understanding of hadron forces. The coupling is too strong for perturbation theory to be used with much confidence, and there is no universal non-perturbative theory available. There seem to be many different fields and many different couplings and force strengths. Hadrons as such do not seem to be elementary objects anyway, but have internal structure. The relationship between the internal and external forces is still only vaguely perceived. Experiment shows that hadron–hadron forces such as those that bind the nucleus are extremely complicated in form and as yet only ad hoc models can be used to describe them.

On the face of it, the forces of nature do not offer much scope for unification. Nevertheless, we do have quantum electrodynamics as a guide to what one really good theory looks like. Also, one might hope that the unification process will cure some of the ills that permeate the other forces. Historically, attempts were first made to unify electromagnetism and gravity, forces that share the common property of being long-ranged, into a combined geometrical theory, but without success. It turns out that a more natural pairing is between electromagnetism and the weak force. Both forces are communicated through a particle with one unit of spin, although the W boson is massive and the photon massless.

To understand the problems that afflict the theory of weak interactions it is necessary to go beyond the lowest order of perturbation and consider the effects of higher-order processes involving the exchange of several quanta. First consider the electromagnetic case. One of the simplest processes involves the scattering of two electrons by the exchange of a single virtual photon. Real scattering events must include the effects of two, three, four, . . . photon exchanges. These correction terms can all be calculated and are smaller by roughly 1/137 for each additional photon. However, there are also the possibilities shown in Fig. 4.8 on p. 127 where the virtual photon emitted by an electron is reabsorbed by the *same* electron. Indeed, this process is possible even for an isolated electron. In general we must sum the effects of all such photon 'loops' involving all numbers of photons being emitted and reabsorbed.

The emission and reabsorption of photons by an isolated electron cannot scatter the electron, but it can endow it with extra energy. When the calculations are performed to find out how much extra energy the answer turns out to be infinite! It is easy to see why. Quantum theory demands that we add together the probabilities of all possible virtual photons being passed round the loop. There is no upper limit on the energy of these. The very energetic photons must be rapidly reabsorbed, of course, but the distance between the ends of the loop can be arbitrarily small. It follows that the total energy flowing round the loop has no known limit. As there is no way to separate the mass-energy of the electron from the energy of its virtual photon cloud, it would appear that the electron must have an infinite mass.

Problems of unlimited self-interaction are not unique to quantum theory. If a classical electron is considered as a little ball of radius r with its electricity spread uniformly across its surface, then the electrical energy *within* the electron due to the repulsion of the self-charge is e^2/r. If the ball is shrunk to a point ($r = 0$) the energy rises without limit. In the quantum case this static self-energy is also present, but it is exacerbated by the emission and reabsorption of photons associated with electromagnetic waves.

Fortunately the problem of infinite self-energy can be overcome. Because there is no way to separate the electron from its electromagnetic field (or photon cloud) then the mass-energy of the electron which we measure must always be the *total* (i.e. its 'bare' mass plus the energy of the electromagnetic field) which is, of course, finite. Mathematically we could think of a 'bare' electron as a hypothetical entity without a field, having a mass which is infinitely negative, so that when the field is switched on the infinite positive energy cancels it out and leaves the finite difference which we observe as the physical mass of the real electron. Manipulating infinite quantities requires some mathematical care, but it can be proved that all observable quantities are finite.

The technique, developed in the 1930s and 1940s, of absorbing infinities into unobservable 'bare' quantities to get a finite answer, is called *renormalization*. It may appear like a trick, and nobody pretends that it is completely satisfactory, but without renormalization the predictive power of quantum electrodynamics would disintegrate. With it, the answers obtained have the legendary accuracy already described.

In quantum electrodynamics not only the electron mass, but also other quantities, such as the electric charge, acquire infinite values. Using renormalization, sensible finite answers can be obtained for all of these quantities. Unfortunately, the presence of infinities prevents the theory from supplying the actual values of the finite remainder.

Although a theory that requires infinite renormalization is suspect, it is at least better than one in which no renormalization is possible, for then the infinities pollute the theory in an uncontrollable and ambiguous way. This is the situation with weak interactions. In fact, infinities occur in weak processes even when they do not occur in the analogous electromagnetic process. Consider, for example, the scattering of a neutron and a neutrino (Fig. 6.10) in the *second* order of perturbation where *two* W particles are exchanged. First a W^- changes the neutron to

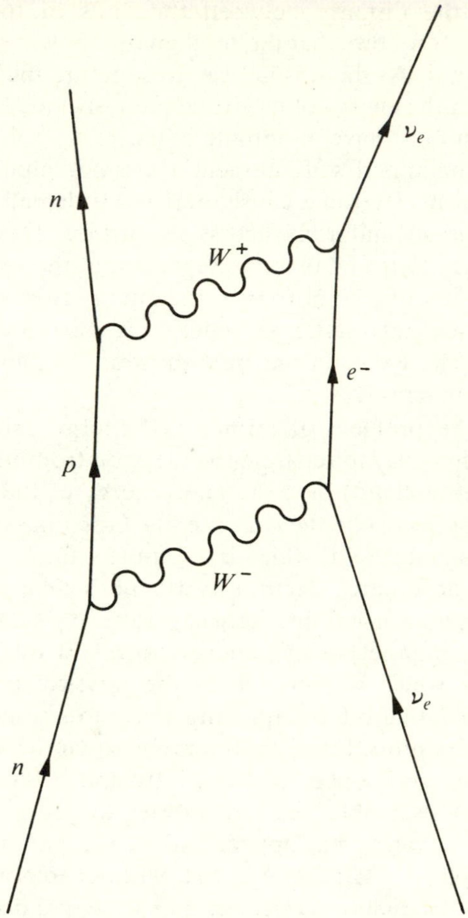

Fig. 6.10. The exchange of two *W* bosons between a neutron and a neutrino appears to give *infinite* quantities instead of a very small correction to the particle motions.

a proton and the neutrino to an electron, then a W^+ exchange restores the original identities. Calculations show that this higher-order process, which should lead to a tiny perturbation, appears to be infinite. However, in this case there is no way to absorb the infinity into a 'bare' quantity. The theory is unrenormalizable.

The renormalizability of quantum electrodynamics can be traced directly to the masslessness of the photon. Like all massless particles that spin, it can direct its spin either parallel or antiparallel to its direction of motion, but not in between as well. In contrast, the massive *W* can align its spin in *three* different directions, for example, parallel, antiparallel and perpendicular to its motion. This seemingly innocuous property is the cause of all the difficulty, because it turns out that it is the *W* particles with the perpendicular spin directions that prevent the infinities from being renormalized away.

These observations suggest that if the *W* particle were massless, it might be possible to construct a unified renormalizable theory of weak and electromagnetic interactions in which the photon and the *W* are combined, like the hadrons, into a single family. Moreover, just as the underlying symmetry of, say, the baryon octet is apparently broken by a strangeness-discerning semi-strong force, so maybe the mass difference between *W* and the photon (admittedly enormously larger than that between the octet of baryons) could be attributed to some sort of internal symmetry breaking.

To understand this important idea it is first necessary to consider both the nature of the internal symmetry, and the way in which it is broken to provide the *W* particle with a large mass. The most familiar symmetries are spatial. Reflection and rotation symmetries were discussed in Section 5.2 and are readily visualizable. We also discussed abstract, or internal symmetries, such as charge reflection and isotopic spin rotation, which, though not visualizable in space, are nevertheless of as much significance in ordering the behaviour of subatomic particles as are the spatial symmetries.

One of the most powerful types of internal symmetries is called a *gauge* symmetry. Only mathematics can provide a proper description of this concept, but a very crude idea can be obtained from one simple example. Recall from Chapter 1 that when an object is lifted from the ground to an elevated location, energy is expended to overcome gravity. The expended energy appears in the object as potential energy, and may be recovered by releasing the object and letting it fall back to the ground. The quantity of potential energy is easily calculated as mass × acceleration of gravity × vertical height lifted. However, we do not really measure the energy of the object by this formula, but only the energy *difference* between the two locations. Moreover, because energy is conserved, it doesn't matter by which path the object is transported from the ground to the elevated place. All that matters is the potential energy difference between the beginning and end of the path.

The freedom to choose any path, or add any fixed quantity of total energy to the system without changing the difference between the two end-points, is a powerful internal symmetry of the system. It is called a gauge symmetry because we can 'regauge' our measurement of energy by any amount, leaving the differences unchanged.

A similar situation occurs for electric forces. A charged particle can be transported between regions of high and low static electric field along any path without changing the difference in electric potential energy between the end-points. In this example, complications occur when moving charges and magnetic fields are permitted, but an extended gauge symmetry exists which encompasses all this. The extended gauge symmetry can be used to deduce all the properties of electromagnetism and is, in particular, deeply associated with the masslessness of the photon. It therefore seems natural to search for a unified theory of weak and electromagnetic forces in which a still more extended gauge symmetry exists – that allows for the incorporation of both photons and the *W* particles. Several such theories have been developed over the last decade, the first being constructed in 1967 by the American physicist Steven Weinberg, and independently by Abdus Salam.

How is the gauge symmetry to be broken to allow the *W* to develop a mass, without destroying the renormalizability of the theory? The answer concerns the phenomenon of *spontaneous* symmetry breaking, which has some familiar counterparts in other branches of physics. The essential feature is that the underlying *forces* remain symmetric, but the actual quantum *state* of the system breaks the symmetry.

A good example of the spontaneous breakdown of symmetry is a bar magnet. The forces acting on the individual atoms of a magnet do not distinguish one direction of space from another, i.e. they are rotationally symmetric. Nevertheless, the magnetized bar obviously defines a preferred direction (N–S). How does an asymmetry which is not present in the forces appear in the collective organization of the bar atoms?

At high temperatures the thermal agitation of the atoms is sufficient to disrupt their collective arrangement and the molecules spin about aimlessly, pointing in no special direction. In this condition the bar has no overall magnetization, and the rotational symmetry of the underlying electromagnetic forces is reflected in this state. The situation can be depicted graphically as in Fig. 6.11*a*. In order to display magnetization, the bar molecules would have to cooperate and spontaneously line up one way, a feat which would actually cost useful energy to organize. As all physical systems naturally seek out the lowest energy state, we see that the condition of zero magnetization, at the bottom of the energy well, is favoured.

At low temperatures the situation changes (see Fig. 6.11*b*). The energy curve adopts a sort of *W* shape which is still symmetric, but in which the mid-point (the point of zero magnetization) is no longer the minimum. The bar then seeks out one of the low points of non-zero magnetization and breaks the symmetry spontaneously. The underlying symmetry is still there, of course, in the fact that the actual direction of magnetization is completely random (a thousand different magnets otherwise free of forces would arrange their magnetic axes a thousand

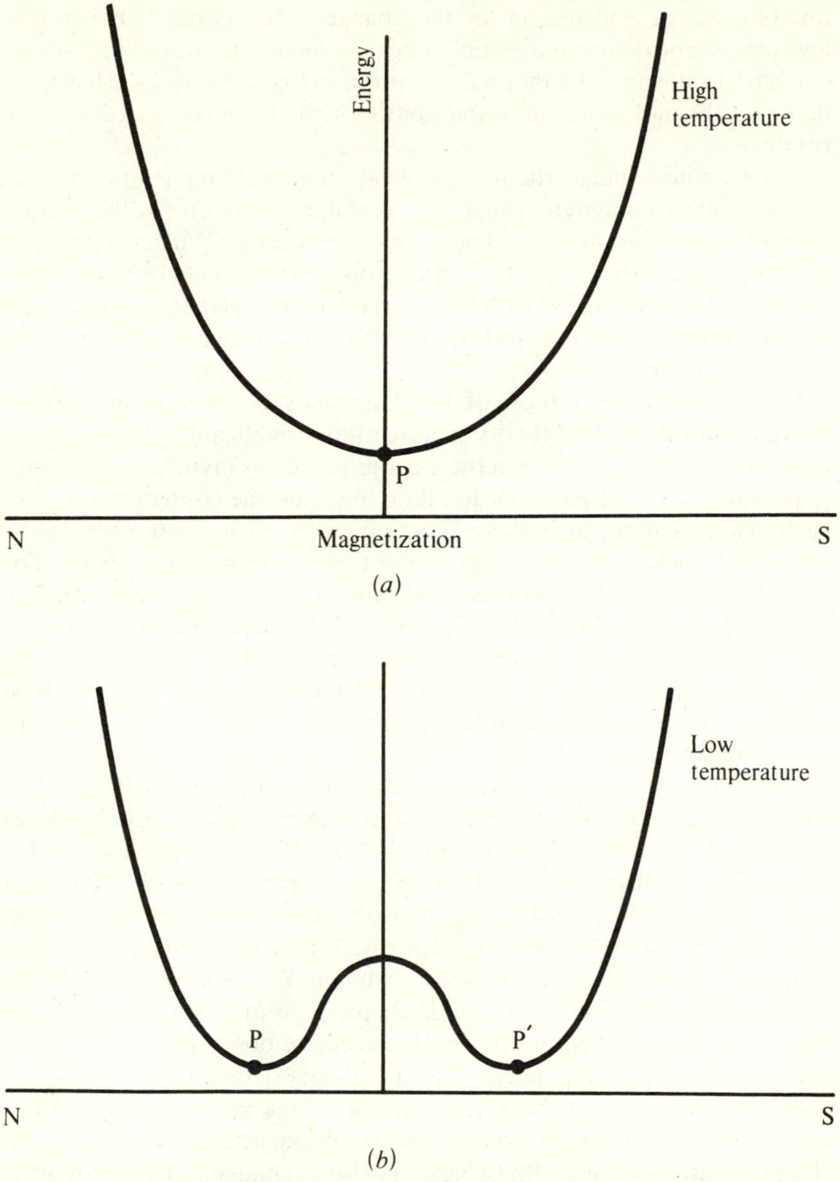

Fig. 6.11. Spontaneous symmetry breaking. A bar magnet will seek out the lowest energy state. In (*a*) this is a point (P) of zero magnetization, which respects the underlying rotational symmetry inside the material of the magnet, because no special magnetic (N–S) axis is defined – there is no magnetism! (*b*) At low temperatures the symmetry is broken, and the bar opts randomly for P or P′ as the equilibrium (lowest energy) state. Whichever state is chosen, it defines a privileged direction of magnetization, but the symmetry is still there, hidden, because the opposite choice could equally well have been made.

different ways). The reason for the change in behaviour is because at low temperatures the magnetic forces between the atoms are strong enough to withstand the thermal agitation, so that the energy released in lining up the molecules more than pays for the increase in order which results.

In the unified gauge theories of weak and electromagnetic interactions, the electromagnetic gauge part remains unbroken (so the photon is massless) but the weak field acquires three massive particles from the spontaneously broken part of the gauge symmetry. Two of these particles can be identified with the charged intermediate bosons W^+ and W^-, but there is also a *neutral* particle, called Z, which was not present in the original Fermi theory.

In 1971 Gerhard 'tHooft of the University of Utrecht in Holland proved that this unified theory is indeed renormalizable. The secret of the renormalizability is that in the various processes involving exchanges of photons, W and Z particles all taken together, the contributions from each, whilst leading to individual infinities, do so in such a way as to cancel all those infinities that cannot be renormalized away. For example, consider the non-renormalizable infinity associated with Fig. 6.10. In the unified theory this diagram must be augmented by processes of the same type in which, for example, neutral Z particles are exchanged instead of the Ws. When all these are added to the type involving W^+W^- exchange, it is found that the remaining infinities are renormalizable.

As a bonus, the theory provides a relation between the weak coupling constant and the masses of the long-sought intermediate particles W, as well as Z. It turns out that W must be at least 39.8 proton masses and Z above 79.6 proton masses. If a W or Z particle is created in a high-energy collision it will decay through the weak interaction in about 10^{-18} s, so it cannot be observed directly. It may be possible, nevertheless, to detect its decay products, for example: $W^+ \rightarrow \mu^+ + \nu_\mu$. Unfortunately, many other processes give decay products of this type, making an unambiguous identification of the intermediate boson difficult.

However, a possible check on the theory can still be made by searching for effects of the Z particle even if the particle itself has not been observed yet. For example, when a W particle is exchanged, the charge it carries ensures that charge is also exchanged. Thus, when an electron scatters from an electron–neutrino, the two are forced to change identity (Fig. 6.12*a*). However, a Z particle allows for so-called neutral currents to exist, where no charge is transferred and the particles retain their identity (Fig. 6.12*b*). A better process to study is the scattering of electrons and *muon*-neutrinos, which can only occur through Z exchange (Fig. 6.13). Several accelerator experiments appear

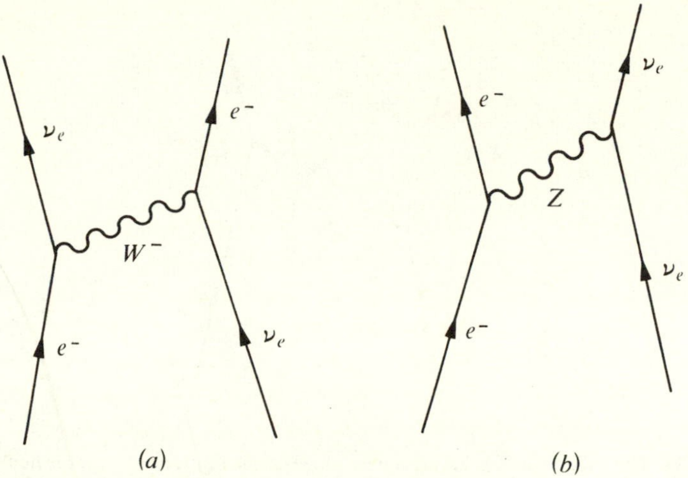

Fig. 6.12. (*a*) When an electron scatters with a neutrino, electric charge can be transferred on the W^- boson, so the two particles e^- and ν_e swap identities. (*b*) Scattering via the neutral Z meson leaves the identities unchanged.

to have detected processes of this type (e.g. $p + \nu \rightarrow p + \nu$), and there is cautious optimism that the existence of neutral currents is being verified.

Another consequence of the unified theory is that the breakdown of parity conservation, so characteristic of the weak interaction, should appear in new, essentially electromagnetic, situations; for example in the behaviour of atomic electrons. In the summer of 1978 the American physicist Richard Taylor reported that he had used the Stanford Linear Accelerator to scatter electrons from protons at high energy. By keeping track of the directions in which the electron spins point, he was able to confirm the existence of a small but vitally significant violation of mirror symmetry.

Through the unified theories of weak and electromagnetic interaction the quarks and the leptons acquire a closer relationship. The combined weak–electromagnetic force sees *four* distinct leptons (e, μ, ν_e, ν_μ) and *four* quarks (u, d, s, c). By absorbing a W boson, neutrinos can be converted into electrons and muons, and quarks can be converted into other flavours. The recent appearance of a *fifth* lepton into this tidy scheme has caused a minor sensation. Several experiments have strongly suggested the existence of a sort of heavy muon, called the τ particle, with a mass of about 1800 MeV. Although a heavy lepton fits naturally into the unified gauge theories, it upsets the correspondence with the quarks, which suggests to many that more quark flavours (and still more leptons?) may exist.

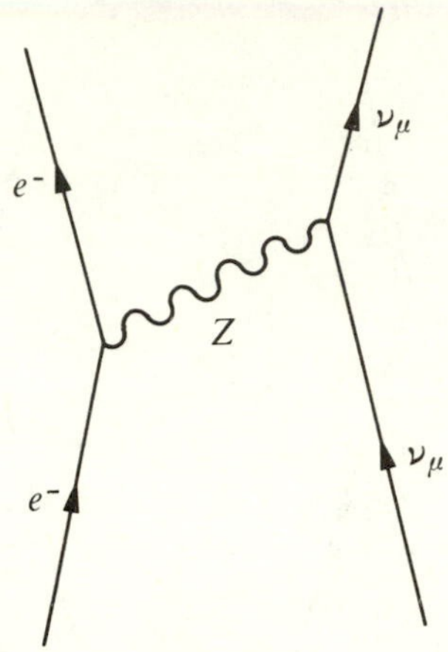

Fig. 6.13. The existence of Z allows a new process: the scattering of *muon*-neutrinos (ν_μ) from electrons.

In the summer of 1977 a new long-lived hadron, called Υ (upsilon), was discovered in proton–proton collision experiments. With a mass of almost 10 000 MeV it seems to be a kind of big brother to the ψ and, like ψ, can also exist in excited states. A natural explanation seems to be that we are observing the antics of yet another pair of quarks, provisionally dubbed 'top' and 'bottom', and that Υ is the analogue of charmonium, i.e. bottomonium. Already the search is on for 'naked bottom'.

There seems to be a natural pairing among the leptons and quarks. Each charged lepton has an associated neutrino, and pairs with two quarks (see Table 6.1). The lightest lepton (the electron) is thus associated with the two lightest quarks (u and d). The next level is like a more massive version of the first, with the heavier muon and its attendant neutrino pairing with the heavier strange and charmed quarks. One can envisage the τ lepton, accompanied by a new τ-neutrino (ν_τ) paired against two new quarks with two new flavours (top and bottom?). Perhaps this sequence is unending?

The success of gauge theories seems to point the way to further unification involving the weak and strong forces also. In this respect it is significant that the general theory of relativity can be cast in a gauge form, and the strong interaction model called quantum chromodynamics (see p. 213) is likewise a gauge theory.

Recently, it has been speculated that at high energies all the forces of nature have comparable strength so are perhaps all merely different faces of a single underlying interaction – a unified gauge symmetric

Table 6.1. Elementary particles? Colourless leptons and coloured quarks could be the building blocks of all matter. They associate together naturally in pairs as shown. The existence of the τ lepton suggests that at least one more level exists. Only the first level seems to be needed to build ordinary matter. (Units: mass in MeV, charge in units of proton charge, spin in units of $\hbar$.)

Leptons				Quarks					
Particle	Mass	Spin	Electric charge	Flavour	Mass	Spin	Electric charge	Strangeness	Charm
Electron-neutrino	0	1/2	0	*d*	300	1/2	− 1/3	0	0
Electron	0.511	1/2	−1	*u*	300	1/2	+ 2/3	0	0
Muon-neutrino	0	1/2	0	*s*	500	1/2	− 1/3	−1	0
Muon	106	1/2	−1	*c*	1500	1/2	+ 2/3	0	+1

theory in which the colour gluons, the graviton, the photon and the *W* and *Z* particles are combined in a grand gauge multiplet. Who knows?

The success of weak and electromagnetic theories stems not just from their gauge symmetric nature, but from the fact that their coupling strength is weak enough to enable perturbation theory to be used, but still strong enough to lead to effects that are measurable in real laboratory experiments. The strong interaction suffers from being too strong to handle using perturbation techniques, whilst gravity is far too weak to have any conceivably observable effects at the quantum level.

In spite of the practical irrelevance of quantum gravity, physicists would very much like to have a good theory, if only for the sake of the completeness and unity of physics. Many attempts have been made in the last 25 years to formulate such a theory, but without success. When gravity is coupled to matter, the result is an endless sequence of non-renormalizable infinities, order by order in the perturbation corrections. Nobody knows how to cure this problem.

6.5 *Conclusions*

In Nature's infinite book of secrecy
A little I can read (William Shakespeare)

The present state of subatomic physics is like a detective story that we have to put down just as the plot is beginning to get exciting. Enough information is available to whet the appetite and suggest that a coherent pattern is starting to emerge, but many burning questions remain to be answered, grey areas to be coloured and pieces of the jig-saw to be fitted.

The two truly monumental achievements in understanding are first the grouping together of different particles into family multiplets, with their compelling interpretation in terms of permanently confined quarks, and second the initial steps towards a unification of all the forces of nature. Throughout, the fundamental link between the structure of matter and the forces which act on it has been apparent.

In contrast, the most conspicuous failure is the lack of a decent theory of quantum gravity, or any real understanding of why it is so deeply separated in strength and behaviour from the other three forces. Not only is gravity absurdly weak, but it operates with a spin-two boson instead of the spin one of the other forces (photon, W boson and gluons). Indeed, this turns out to be the very reason why gravity has a *geometrical* interpretation (a spin-one particle does not have enough geometrical structure to describe curvature).

Another conspicuous failure is the inability of mathematics to provide the *numbers* which enter into the theory of the forces. First and foremost are the particle masses. Apart from the limited success in predicting the W and Z masses from the unified gauge theory, it seems to be necessary to put in the particle masses by hand. A complete theory ought to be able to provide not only the multiplet mass splittings, such as that between the proton and neutron, but the masses of all the subatomic particles themselves. This is especially true of the leptons, which are not troubled by the strong interactions. In particular, we should like to know whether there is any significance in the fact that the muon, pion and kaon masses are almost exactly $3/2 \times 137$, 2×137 and 7×137 electron masses, respectively. Is there any connection with the fine structure constant 1/137, and where does this universal number come from anyway?

Any full theory must explain not only the strength of the electromagnetic coupling, $e^2/\hbar c$, but also why the strengths of the other forces are so different. We would like to know why there are so few charge states among subatomic particles. Why are no particles found with charges of $9e$ or $52e$, for example? Similarly with the other quantum labels like strangeness or charm, why are the limits so narrow?

Great mystery also surrounds the conservation laws. Why are some, such as energy and lepton number conservation, obeyed by all the forces (with the possible exception of gravity) while others, such as strangeness conservation, violated by the weak force? How many 'invisible' labels, such as colour, are needed for the complete set of laws? What is the meaning of the non-dynamical conservation laws that cannot be connected to space–time symmetries?

There are also shortcomings in the experimental situation. What about the particles that should be there, but aren't, such as the *W* boson, the *Z* boson and the magnetic monopole, not to mention the particles that some believe are there but may never be seen, such as the quarks and gluons?

Quite apart from problems with theory and experiment, there are deep philosophical questions about the structure of subatomic matter. Thinking back over the particles described in this book, and remembering the many that were omitted or given only a passing mention, one is bound to be struck by the thought: what are they all *for*? The tremendous superfluity of species seems rather like having characters in a play in which the principle actors number only a handful among a vast multitude. Why are all the extras needed?

It is a remarkable thought that as far as we know the world would be much the same place if there were only the *two* lightest quarks (u, d) and the *two* light leptons (e, ν_e). These particles apparently make up all of ordinary matter. A restriction to these elementary constituents would limit the total number of subatomic particles to just a handful – protons, neutrons, electrons and a few more. There seems to be no good reason why we need strange or charmed particles (a result of having the heavier s and c quarks) or the corresponding heavier muons, and the muon-neutrino. It just looks like a duplication of effort. The profligacy is even more astonishing if there turns out to be triplication, as the existence of the still heavier τ lepton and Υ hadron seems to imply. How extended is this overprovision? How many quarks and leptons will there turn out to be? Perhaps there is no limit, and progressively higher energies will uncover ever more massive pairs of associated quarks and leptons, presumably with new characterizing labels analogous to strangeness and charm?

One reason for the superabundance of species might be that nature produces everything it *can* rather than everything it *needs*. We already know from the reactions and transmutations discussed in the previous chapters that every process that can happen does, unless some law forbids it. Maybe the same applies to the production of particle species. On the other hand, if this is so, what has become of the magnetic monopoles, not to mention the bizarre, but entirely possible, faster-than-light particles called tachyons? Is it only experimental inadequacy that prevents their detection?

If nature is in the habit of overproducing particles, it does not seem to have repeated the tendency with the forces themselves. Without gravity there would be no galaxies, stars or planets. Without electromagnetism there would be no heat, light, atoms or chemistry. Without the strong force there would be no atomic nuclei, the world would be made of hydrogen, and the sun and stars could not burn. Life would be impossible under any of these conditions. Even the weak interaction, although it may perhaps be regarded now as just a part of a unified weak and electromagnetic interaction, is probably vital for life as we know it, by controlling the nuclear reactions in the sun and by providing the force which blasts stars apart with neutrinos in so-called supernovae explosions. These events spread biologically vital elements such as carbon, which are synthesized in the stellar interiors, into interstellar space, where they can eventually be incorporated into the material of planets like the Earth, and thence into biological organisms.

The investigation of the forces of nature described in this book has been based on the underlying philosophy that by pulling matter into ever smaller pieces the true elementary structure will be exposed. In a limited sense this expectation has been justified. First, the study of molecules led to a good understanding of the properties of gases and solids, then breaking molecules into atoms revealed the properties of chemical bonds. Stripping electrons from atoms showed where electricity comes from and how light can be emitted and absorbed, while smashing the nucleus itself uncovered totally new forces. Finally, breaking open the nucleons suggests a still more elementary level of structure and even more forces.

We might call this approach *analytic* physics. However, nature has another side to it – the synthetic. We would not expect to investigate how a clock works by smashing it into bits and pieces with a hammer and studying the cogwheels in isolation. The feature that distinguishes a clock from just a heap of cogs and springs is its *collective organization*. A clock is a cooperative mechanism. In the same way, we cannot expect to understand life by pulling DNA apart. Searching for the laws of biology among atoms of carbon and oxygen would be futile. There seems to be a kind of uncertainty principle which says that one cannot simultaneously determine how matter *works* and what it is *made of*. The very act of investigating its structure inevitably destroys its organization. Analytic physics can at best be only half the story.

Very little has been said about the synthetic properties of matter in this book. A short discussion was given about how the properties of gases can be derived from the accumulated behaviour of billions of molecules, and about how the asymmetry in time of macroscopic systems is manifested through a one-way disintegration of their collective organization. Mostly we have concentrated on analytic probing into progressively smaller and less accessible recesses of matter. But how far can the analytic approach be taken? Has it already foundered at the level of quarks that cannot, even in principle, be studied in isolation, but whose properties are only manifested through their cooperative behaviour?

The expectation that a greater expenditure of money and energy will enable ever simpler constituents of matter to be smashed into existence seems threatened by the prospect that endless quark–lepton duplication will be the only thing to show for the effort. Could it be that our present level of understanding is a half-way stage – an island of simplicity – between macroscopic and submicroscopic complexity? The bootstrap theory is the clearest example of how it could be that on a very small scale the forces and particles might be controlled by the same kind of collective and cooperative order that characterizes complex macroscopic matter. If this is so, then it may be necessary to ask completely different questions than hitherto asked about the organization of the microworld. Instead of wondering how matter is put together, we need to study the forces and fields as an integrated whole. There is a risk that we may miss the picture on the jig-saw because of a preoccupation with the shapes of the pieces and the way they fit together. In the seventeenth century mankind faced a revolution in perspective when it was realized how insignificant and diminutive the Earth is in relation to the vastness of the universe. This century we have looked inwards to the secret depths of matter, and marvelled at another revolution in perspective. The apparently solid matter of our senses has given way to an insubstantial world of transitory particles and the shifting complexity of their unending interactions; a world in which strange laws of conservation compete against quantum randomness and chance to prevent undisciplined descent into chaos and annihilation.

Undoubtedly the most awesome fact to emerge from our study of the microworld is how diabolically clever nature turns out to be. Many of the discoveries described in this book are the result of intense and collective intellectual activity by some of the finest minds the world has ever known. Some of the theories have called upon subtle and obscure mathematical arguments that could easily have been overlooked, even by highly competent mathematicians. Yet nature has been smart enough to spot them: to build up multiplets from SU(3) symmetry groups, to use the simplest and most beautiful gauge symmetry to construct electromagnetism, and to spot the loophole in the mathematical theorem that might have prevented the *W* boson acquiring a mass. Mathematics and beauty are the foundation stones of the universe. No one who has studied the forces of nature can doubt that the world about us is a manifestation of something very, very clever indeed.

Index